Steven A. Benner (Ed.)

REDESIGNING THE MOLECULES OF LIFE

Springer-Verlag

REDESIGNING THE MOLECULES OF LIFE

Editor:
Prof. Dr. Steven A. Benner
ETH Zürich
Laboratory for Organic Chemistry
Federal Institute of Technology (ETH)
CH-8092 Zürich

REDESIGNING THE MOLECULES OF LIFE

Conference Papers of the International Symposium
on Bioorganic Chemistry
Interlaken, May 4–6, 1988

Sponsored and Organized by
The Association of Swiss Chemists

Springer-Verlag
Berlin Heidelberg New York
London Paris Tokyo

Prof. Dr. Steven A. Benner
ETH Zürich
Laboratory for Organic Chemistry
Federal Institute of Technology (ETH)
CH-8092 Zürich

QP
517
.585
I58
1988

ISBN 3-540-19166-6 Springer-Verlag Berlin Heidelberg New York
ISBN 0-387-19166-6 Springer-Verlag New York Berlin Heidelberg

Printing and Binding: Weihert-Druck GmbH, Darmstadt
2151/3140-543210

PREFACE

ιe organic chemist is rarely satisfied by a simple "explanation"
: the reactivity of organic molecules. Rather, the chemist
ιnts to go one step further, to "control" the behavior of
ɔlecules by altering their structure in a controlled way. This
s, in fact, a rather stringent definition of "understanding," as
: requires the "prediction" of behavior from structure (or
:ructure from behavior). But it also places technical demands
ι the chemist. He must be able to synthesize the molecules he
:udies, characterize them at the atomic level of structural
:solution, and then measure their behaviors to the precision
ιat his explanation demands.

.ological chemistry presents special problems in this regard.
.though the tools for synthesis, purification, and structural
ιaracterization are now available for manipulating rather large
.ological macromolecules (proteins and nucleic acids in
ιrticular), the theory supporting these manipulations is
ιadequate. We certainly do not know enough to control generally
ιe behavior of biological macromolecules; still worse, it is not
.ear that we know enough to design synthetic molecules to expand
ιr understanding about how reactivity in such biological
.cromolecules might be controlled. Starting from scratch, there
ι simply too many oligopeptides to make; starting from native
ʼoteins, there are simply too many structural mutations that might
ι introduced.

ιis paperback is the conference documentation of a symposium
ιtitled "Redesigning Life," to be held in Interlaken from May 4 to
 1988. The manuscripts span the range chemistry to molecular
ology. Each speaker has been instrumental in developing
ιproaches for solving the problems inherent in complexity of
ological macromolecules. Richard Kellogg has abstracted the
ιtails of biochemical reactivity, and rebuilt these into designed
ʼganic "model systems." Alan Fersht has shown how a sufficiently

A 9|7|88

large collection of structural variations in a natural enzyme can, if supplemented with rigorous physical organic analysis, yield a sophisticated picture of the origin of catalysis. Tom Kaiser has used a chemical understanding to bridge the gap between structure and biological activity of small polypeptides, using abstractions of natural polypeptides to design synthetic polypeptides that mimic their biological activity. Jack Szostak is showing how a molecular biological approach enables the scientist to study and perhaps recreate catalytic activity in RNA molecules. Finally, I have provided a chapter outlining the construction of historical and functional models explaining the peculiarities in the behavior of biological macromolecules.

The major contributors generously prepared manuscripts several months in advance so that this book could be in the hands of the participants at the time of the symposium. This has one major disadvantage. In a field that is moving as rapidly as this one, exciting results will undoubtedly emerge between now and May that will be discussed at the conference but could not be presented here. I hope that in spite of this deficiency, the book will assist scientists on both sides of the "great divide" separating chemistry and biochemistry to bridge the gap, and help to usher in the age where research ideas and methods will be applicable in a continuum of chemical problems that range from the smallest molecule to the largest organism.

Zurich, 31 January, 1988   Steven A. Benner

CONTENTS

NICOTINAMIDE AS A COENZYME. AN EVOLUTIONARY MODEL OF CATALYSIS AND SOME ATTEMPTS TO TRANSLATE IDEAS INTO TESTABLE CHEMISTRY

Richard M. Kellogg[#] and Cornelis M. Visser[†]

\# Department of Organic Chemistry, University of Groningen
   Nijenborgh 16, Groningen 9747 AG, The Netherlands
† Open University, Science Department
   (contact address) Jouwer 6, Sebaldeburen 9862 TA, The Netherlands

Introduction

In this article we shall summarize some of our postulations, experiments, and some of the questions that have arisen during our work on catalysis by nicotinamide and related questions. One of us (C. M. V.) has approached bio-organic chemistry from an evolutionary standpoint. He has tried to devise a consistent picture of how bio-catalysis might have developed derived, at least in part, from analysis of possible relics of prebiotic chemistry in modern metabolism.

R. M. K. has used a classical chemical approach. Inspired by the possibilities of crown ethers to provide designed binding sites, he has synthesized molecules that, at least at first glance, appear to imitate some of the basic chemistry involved in dehydrogenases. The goal has been, and remains, to develop catalytically active systems.

These approaches are widely (perhaps even wildly) divergent. We believe that they are complementary; we have in the effort to pick up these threads of complementarity often been led to new insights and ideas.

Frontier Orbitals and Enzymatic Catalysis as Related to NAD(P)H.

NADH and NADPH regulate group transfer.[1] The group that is transferred does not exist as such in solution. The coenzyme ATP, for instance, transfers the phosphoryl group, PO_3^- , which in water is bound to a molecule of water. The phosphoryl group has the character of a reactive intermediate like a carbonium ion. function of the coenzyme is to bind the group that has to be transferred in a chemical bond that under most circumstances is kinetically stable, but which can become kinetically reactive under defined catalytic conditions.

S. A. Benner (Ed.)
Redesigning the Molecules of Life
© Springer-Verlag Berlin Heidelberg 1988

The group transferred by NADH and NADPH is the hydride ion, instable in water like the phosphoryl group. This hydride is bound in NAD(P)H in the form of a thermodynamically usually relatively stable C-H bond. With the help of very simpl theory[2] we try to understand by what means this C-H bond can be activated, and why this reaction has been selected very early in evolution as one of the important redox reactions.

NADH and NADPH play different roles in most organisms. They take part in different reaction systems: NADH in catabolic, energy supplying-, NADPH in anabolic, energy demanding reaction sequences. There is, however, always at least one enzyme, transhydrogenase, that regulates the ratio of the two redox couples by catalyzing the hydride exchange reaction between the two systems:

$$\text{NADH} + \text{NADP}^+ \rightleftharpoons \text{NAD}^+ + \text{NADPH}$$

We choose the almost symmetrical reaction for a closer qualitative molecular orbital description. Although we restrict ourselves to reactions with a positivel (and stabilized) hydride acceptor (the formal carbocations NADP^+ and NAD^+), simila considerations are applicable to enzymic reduction of neutral carbonyl groups. Neutral carbonyls like that of acetaldehyde (alcohol dehydrogenase) and pyruvate (lactate dehydrogenase) are bound at the respective active sites with the help of zinc(II) ion or a histidinium ion (fig 1). The hydride acceptor site, the carbony carbon complexed to these electrophilic centers, can also be considered as a forma carbocation.

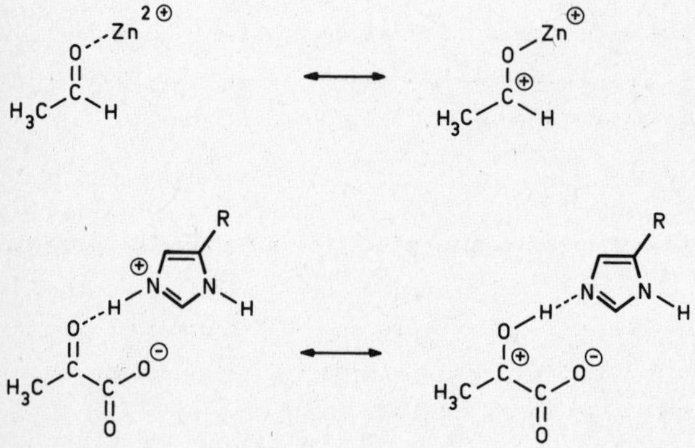

Figure 1. The formally cationic hydride acceptors on the active sites of a. alcoho dehydrogenase and b. lactate dehydrogenase.

Hydride transfer from 1,4-dihydropyridines is now thought, after long debate,
be a one step reaction during which no unpaired electrons arise unless the
action partner strongly demands a radical intermediate as in the reduction of a
rri to a ferro ion.[3] There seems to be some, but not overwhelming, bias for a
near transition state.[4] Apart from structural variation of the transition state
ere are in addition several possibilities for the charge distribution. Three
tremes are give in Fig 2. The question of the charge on the moving hydrogen is
portant but difficult to answer.

$$\text{a)} \quad C-H \;+\; \overset{\oplus}{C} \;\longrightarrow\; \left[\; \overset{\oplus}{C}----H^{\ominus}----\overset{\oplus}{C}\;\right]^{\ddagger}$$

$$\text{b)} \quad C-H \;+\; \overset{\oplus}{C} \;\longrightarrow\; \left[\; \overset{\oplus\frac{1}{2}}{C}--H----\overset{\oplus\frac{1}{2}}{C}\;\right]^{\ddagger}$$

$$\text{c)} \quad C-H \;+\; \overset{\oplus}{C} \;\longrightarrow\; \left[\; C----H^{\oplus}----C\;\right]^{\ddagger}$$

gure 2. Three extremes for a symmetrical charge distribution in the transition
       state for hydride transfer.
       a. hydride ion: maximal negative charge on H
       b. hydrogen atom: minimal (zero) charge on H
       c. proton: maximal positive charge on H.

perimentally it has been investigated by estimation of the charge on both carbon
oms via a Hammett approach but this method is not very reliable.[3]

Theoretical calculations give varying results but in all cases the charge on
e moving hydrogen is small. This question is an important one since the charge
the transition state hydrogen may determine the way in which this transition
ate can be stabilized through an interaction with the catalyst. If the charge is
gative, a redox neutral Lewis acid like a zinc ion could possibly give
abilizing interaction. But if the charge on the moving hydrogen is slightly
sitive then a hydrogen bridge-like interaction could be the origin of catalysis,
ovided the H-bridge acceptor is not basic and will not become protonated. An
ide oxygen atom would be a good candidate for such a stabilizing interaction.

The transition state of a symmetrical hydride transfer reaction, like the
action catalyzed by transhydrogenase, consists of a hydrogen nucleus surrounded
two carbon atoms at the same distance. From other areas of chemistry we know two
pes of similar - but stable - structures. One is a symmetrical hydrogen bridge
ke the strong one found in the FHF⁻ anion. Linear FHF⁻ can be described as a
oton surrounded by two fluoride anions at a distance of 113 pm, which is 54 pm
orter than two times Van der Waals radius of a fluoride

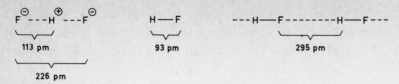

113 pm

93 pm

295 pm

226 pm

Figure 3. Bond length in the FHF$^{\ominus}$ anion compared with those in HF.

ion (Fig. 3).[5] This hydrogen bond is mainly ionogenic but there is a covalent contribution as well. The schematic bonding scheme for this situation is given in Fig 4. For the case of FHF^{-} the highest occupied orbital (HOMO) is the non-bonding MO.

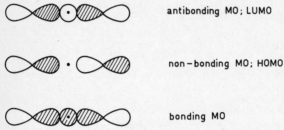

antibonding MO; LUMO

non-bonding MO; HOMO

bonding MO

Figure 4. Simple three atom MO scheme. For HFH^{-} the bonding and non-bonding orbitals are occupied; the latter is the HOMO. For the 1,5-cyclo-decyldication there are only two electrons; the bonding now becomes the HOMO.

The other type of relatively stable hydrogen atom surrounded by two heavier nuclei is the 1,5-hydridocyclodecyl carbocation described by Sorensen, et. al.[6] This is illustrated in Fig. 5. The bridgehead carbon atoms are somewhere between

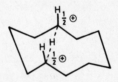

Figure 5. The 1,5-hydridocyclodecyl carbocation.

sp^2 and sp^3 hybridized. For a qualitative description of the hydrido bond we can take them as being sp^2 hybridized, and can consider the cation to originate from the combination of a 1,5-cyclodecyldication and a hydride ion. NMR spectroscopy of hydrido-bridged cations has shown that the negative charge on hydrogen is in fact much lower than 1: only ca 0.1 electron, in agreement with the small difference in

ectronegativity between carbon and hydrogen. The C--H--C bond is therefore
ainly covalent. The bond energy of each C--H bond is ca 220 kJ mol^{-1}, so that the
nergy of the cation with hydrido-bridge is only ca 20 kJ mol^{-1} lower than that of
e same cation without a C--H--C bond.[6b] A qualitative description of the C--H--C
ond is obtained by linear combination of two unfilled $2p_z$ AO's of the carbon atoms
d one 1s AO of a hydride ion filled with two electrons. This gives a MO bonding
heme similar to that for the FHF$^-$ anion, but with a smaller energy difference
etween the MO's and now with only two electrons so that now the bonding MO is the
MO and the non-bonding MO is the lowest unoccupied one (LUMO). Refer again to
g. 4.

The structure of the transition state of the non-catalyzed transhydrogenase
eaction can be approached in the same manner as the structure of the 1,5-
dridocyclodecyl carbocation. The hybridization of the two carbon atoms is again
etween sp^3 and sp^2 but we use again the LUMO's of the two (sp^2 hybridized)
ridinium cations[7] for linear combination with a (filled) 1s AO of hydrogen,. In
ese LUMO's the coefficient on C4 is the highest; C4 is the soft site of the
ectrophilic pyridinium ion and attack by a soft hydride transferring organic
olecule occurs preferentially at this site. Relatively hard hydride anions like
ose delivered by LiAlH$_4$ have some preference for the other positively charged
arbon atoms C2 and C6.[8] An MO bonding scheme for the transition state is obtained
at is essentially similar to that in Fig. 4, in which the 2pz AO's are used in
ace of the LUMO of pyridinium ions. The bonding MO in Fig. 4 corresponds with
e highest occupied MO for hydride transfer, the HOMO$^{\#}$, and the non-bonding MO
th the LUMO$^{\#}$. These frontier orbitals provide a qualitative description of the
ansition state, and, most importantly, they can be used to understand the ways in
ich the transition state can be specifically stabilized, in other words how
atalysis might be obtained.

MO calculations on the structure of hydrido-bridged carbocations give a C--C
stance of about 260 pm.[6a] For the transition state for hydride transfer one
nds on the average values of about 290 pm.[9] This difference reflects the
stinction between a stable intramolecular hydrido-bridge and an unstable
termolecular hydrido-bridge during a transition state. The bonding MO of the
drido-bridge is more extended in the transition state and the overlap between the
 AO of hydrogen and the two LUMO's must by consequence be considerably smaller
mpared with the stable hydrido-bridged cations, since the accumulation of
ectron density is primarily caused by this overlap. Although some calculations
dicate at the transition state a small negative charge on hydrogen of ca 0.1
ectrons,[10] the results of calculations are far from unanimous. Some do find

indeed a <u>positive</u> charge on the migrating hydrogen atom.[11] It seems therefore wi
in any case to reckon with the possibility that in the diffuse transition state t
"hydride ion" being transferred may actually bear a small positive charge. In su
a case stabilization might be achieved by formation of a hydrogen bond with for
instance a carbonyl group. Since the presumed positive charge on the migrating
hydrogen atom exists only in the transition state and not in substrate and produc
this stabilizing hydrogen bridge would certainly be a potential source of
catalysis.

The carboxamide group of nicotinamide, so far ignored in this analysis,
rotates about the carbonyl carbon-ring bond with only a low energy barrier. In X
ray diffraction structures of NAD(P)$^+$, of NAD(P)H analogs and models and of other
nicotinamides this substituent is found in all possible orientations.[12] Buck, et
al.[13,14] have calculated that the activation energy for hydride transfer should b
lower if the carbonyl oxygen points towards the migrating hydrogen atom so that t
distance between the oxygen atom and hydrogen atom being transferred is minimal.

Qualitative MO and perturbation theory (the Klopman-Salem equation)[15] offers
two possible explanations for this possible catalytic effect of a carboxamide
group. One is a favourable frontier orbital interaction and the other a favourab
Coulomb effect. Consider the first possibility. The local frontier orbitals of
the carboxamide group are qualitatively similar to those of an allyl anion as
illustrated in Fig. 6.[7] The HOMO and LUMO of formamide are taken as models. Whe
the distance between the oxygen atom and the migrating hydrogen atom is minimal

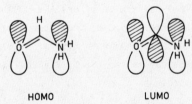

HOMO   LUMO

Figure 6. HOMO and LUMO of formamide, used as models for the local frontier
          orbitals of the carboxamide group.

corresponding to a dihedral angle between carboxamide group and pyridine ring of
60° there is a stabilizing HOMO$^\#$/LUMO$_{cat}$ interaction present that is absent in th
beginning situation. It is, however, only a weak overlap between one lobe of the
AO on oxygen and migrating hydrogen atom, see Fig. 7. The other frontier orbital
interaction, LUMO$^\#$/HOMO$_{cat}$, is absent because the coefficient on the moving
hydrogen

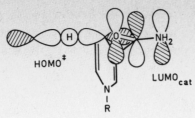

Figure 7. Weak HOMO$^{\#}$/LUMO$_{cat}$ interaction during hydride transfer. Of the HOMO$^{\#}$ only the AO's of C-4 atoms and hydrogen are shown.

is zero (see Fig. 5). The total contribution of the frontier orbital term to stabilization of the transition state is likely too small to explain the calculated lowering of the energy of 15 kJ mol^{-1}.[13,14] Additional stabilization may be present. The first term of the Klopman-Salem perturbation equation is chiefly an electrostatic one. If the moving hydrogen atom has a small positive charge this would aid also in stabilization of a hydrogen bridge with oxygen of the amide. These interactions are not present in the initial situation, so that the total interaction would contribute to stabilization of the transition state, i.e. catalysis.

During the last decennium the structures of a number of enzyme/substrate and enzyme/product complexes have been determined at the atomic level by X-ray crystallography.[16,17] Since these structures do not differ as far as the main chain conformation is concerned, one assumes that the main chain structure does not change appreciably during the reaction, in contrast to what happens during binding of substrates and release of products. Since the structure of the main chain apparently remains fairly invariant during the reaction we can get a good impression of the global structure of the transition state by examining enzyme/substrate and enzyme/product structures. If the examination is restricted to the X-ray structures of NAD(P)H-dependent dehydrogenases we find that most enzymes hold the carboxamide group indeed in the right position, namely that in which the amide oxygen and the migrating hydrogen would be in closest proximity, in agreement with the catalytic effect of the carboxamide group postulated by calculation. This preference in the conformation of the carboxamide group is not found in the structures of free coenzymes and models thereof. Amoung the enzymic systems there are, however, some exceptions to this generalization.

One such exception is lactate dehydrogenase from pig heart. In this enzyme there is, however, an asparaginyl side chain present at the place where the hydrogen atom is in the transition state.[16] The orientation of the asparaginyl carboxamide group is not subject to the same limitations as that of nicotinamide, which only can rotate about the sigma bond axis to the heterocyclic ring. If the

orientation is such that the frontier orbitals of the catalyst with two lobes of
the AO of the oxygen atom interact with the frontier orbitals of the transition
state (see Fig. 8) then a stabilizing LUMO/HOMO$_{cat}$ interaction exists but at the

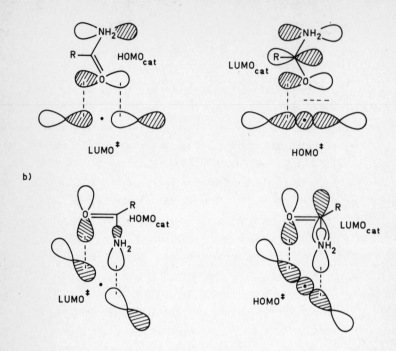

Figure 8. Frontier orbital interactions between the asparaginyl carboxamide group
and the transition state for hydride transfer;
a. interaction only via AO of oxygen
b. interaction via AO's of oxygen and nitrogen

same time a destabilizing HOMO$^{\#}$/LUMO$_{cat}$ interaction is generated. The frontier
orbital term of the perturbation equation is small in this orientation but the
interaction between a positively charged hydrogen and the amide dipole would be
optimal. Other orientations that lead to potential stabilization can also be
devised.[17]

Let us now consider a possible case of stabilization of the transition state
by an amino group. Glutathione reductase catalyzes the reduction of glutathione
disulfide (GS-SG) into two molecules alanylcysteinylglycine (GSH) with the help of
NADPH.

$$\text{GS-SG} \quad + \quad \text{NADPH} \quad + \quad \text{H}^+ \quad \rightleftharpoons \quad 2 \text{ GSH} \quad + \quad \text{NADP}^+$$

e enzyme contains FAD as a prosthetic group. FAD is reduced during the reaction
d transfers both electrons of hydride, but not the proton, to an active site
stine. The two cysteinyl residues thus formed pass the reduction equivalents on
glutathione in a disulfide/dithiol exchange reaction.[18] We limit ourselves to
e first step, hydride tranfer from NADPH to FAD. Glutathione reductase belongs
the group of enzymes that bind the carboxamide group of NADPH in the "wrong"
shion, i.e. oxygen away from the migrating hydrogen.[19] A potential catalytic
oup is therefore expected at the place of the hydrogen atom just as with lactate
hydrogenase. The X-ray crystal structure of the enzyme/NADPH complex shows that
ere is not a carboxamide group present but rather the amino group of a lysyl
sidue, held in position by a glutamyl residue via two hydrogen bonds as shown in
g. 9. Under normal conditions (pH 7 in water) the amino group is completely

gure 9. Lysyl-glutamyl couple at the active site of glutathione reductase. The
terminal nitrogen atom of lysyl is situated at the place of hydride
transfer, half way between NADPH and FAD.

otonated and one assumes that most lysyl residues in and on proteins also bear a
sitive charge. The lysyl amino group is, however, situated at the "positive end"
an alpha-helix, in a strong dipole field that destabilizes a positive charge
preciably.[20] Therefore we may assume that the pK_a of this group lies below the
rmal value of 10.5 and that it can easily be deprotonated (and reprotonated). In
g. 10 the two frontier orbital interactions between the local HOMO of the $-CH_2NH_2$
oup and LUMO$^\#$ as well as the interaction between the local LUMO and HOMO$^\#$ are
etched. The HOMO$_{cat}$ contains essentially the lone pair of nitrogen; the LUMO$_{cat}$
in fact the sigma star$_{C-N}$ orbital. Compare the HOMO and LUMO of methyl amine.[7]

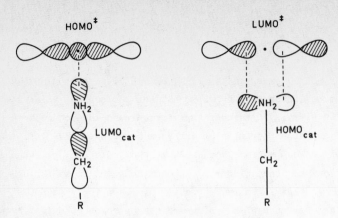

Figure 10. Frontier orbital interactions between the terminal amino group of lysi
and the transition state for hydride transfer.

From Fig. 10 it can be seen that now both frontier orbital interactions are
stabilizing provided that the C-N bond is directed towards the hydrogen atom in t
transition state. This is at the same time a good orientation for a hydrogen
bridge between a weakly negatively charged nitrogen atom and a transition state
containing a positively charged hydrogen atom. In the crystal structure of the
enzyme/NADPH complex[19] the C-N bond has not yet the optimal orientation for
catalysis. It points exactly beside the hydrogen atom that has to be transferred
The lysyl side chain, however, does not have the normal all trans zigzag
conformation of a saturated aliphatic carbon chain. On the contrary all three CH
CH_2 bonds are in a gauche conformation. Manipulation with models shows that such
side chain can easily assume other conformations in which the C-N bond is oriente
toward the migrating hydrogen atom. The double hydrogen bridge with the glutamyl
side chain (Fig. 9) ensures that the nitrogen atom stays in the right position
during this assumed conformational flip. Although the lysyl amino group in this
mechanism leads the hydrogen atom in the direction of flavin, the proton does not
arrive at flavin. The steric hindrance between lysyl and flavin is such that the
is no place for an extra proton on flavin.[18] The two electrons of hydride go
through to flavin (and via flavin to the active site cysteine) but the proton sta
at the same side of flavin and goes in all likelihood to the amino group of lysyl
which thus acts as a base. That a base like lysyl is capable of deprotonation of
transition state for hydride transfer would be a strong argument for a positively
charged hydrogen atom in that transition state.

A Brief Synopsis of Some Results with Model 1,4-Dihydropyridines.

   The simple picture that been sketched so far rests on the presumption that
ring the transition state for "hydride" transfer the hydrogen atom has a modicum
 positive charge. This transition state might be stabilized by hydrogen bond
rmation with the oxygen of the carboxamide group or with other suitably placed
osthetic groups. To our knowledge this point has not been subjected to
ambiguous experimental test in a model system. We will discuss shortly how this
ght possibly might be done.

   Many more factors are involved in enzymic catalysis, and some of these lend
emselves more readily to experimental test than those discussed thusfar.
nsider alcohol dehydrogenase (Fig. 11), which has a catalytically active zinc
om at the active site.[21] One role of the zinc ion is doubtless to polarize the
rbonyl group making the carbonyl carbon electron deficient (positively charged in
 extreme canonical form) and a better hydride acceptor. We suspected also that
e unusual coordination about the zinc ion - two cysteines and a histidine - is
portant in generating the amphoteric character that the ion must achieve to
talyze reaction in both directions.

   This supposition is supported by experiment. These experiments require,
wever, some special approaches. Zinc thiolates are in general oligomeric or
lymeric because of the tendency of thiolate to act as a bridging ligand between
o zinc ions (Fig. 11).[22,23] In alcohol dehydrogenase the rigid structure of the
zyme ensures that only the monomeric zinc complex can be present. Model systems
n be synthesized that are monomeric if the ligands used to coordinate zinc are
idendate and sterically hindered enough to prevent oligomerization. Two examples

gure 11. Enzymatic transition state for alcohol dehydrogenase.

are shown in Figs. 12 and 13. In Fig. 12 a cobalt containing ligand is
illustrated; the corresponding zinc complex is apparently oligomeric despite

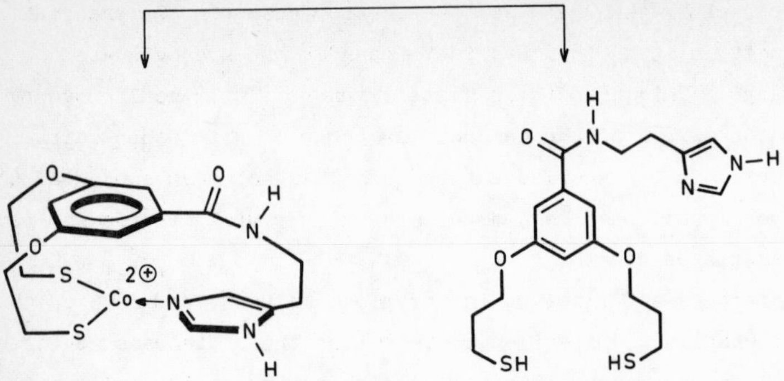

Figure 12. Tridentate ligand designed to encapsulate metal ion. On the right the
structure of the presumed cobalt complex is illustrated.

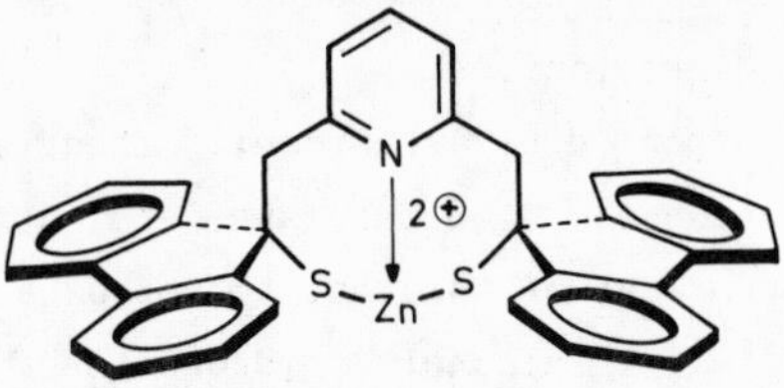

Figure 13. Structure of monomeric tridentate zinc thiolate complex.

the tridentate nature of the ligand. Cobalt(II) ions have, however, been shown t
be excellent surrogates for zinc(II) in alcohol dehydrogenase.[24] Indeed, this
cobalt complex is capable of carrying out in a <u>catalytic</u> fashion (each molecule o
complex functions 100-200 times) in the model reduction depicted in Fig. 14. The
proton source is usually a trace of water, although proton donors like NH_4Cl or
CH_3CO_2H may also be used. The extremely poor solubility of the complex limits it
effectiveness, however.

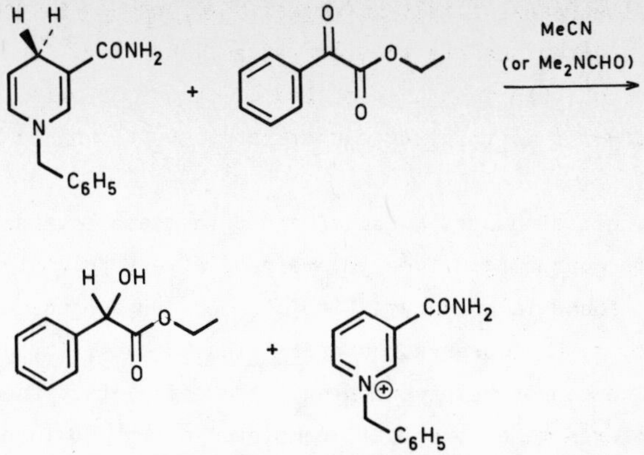

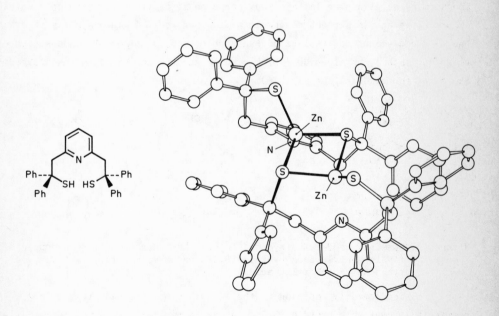

Figure 14. Reduction mediated by the cobalt complex illustrated in Fig. 12. The complex is present in 10^{-4} M concentration and the other components at 10^{-1} M. The turnover on the complex is 100-200 before exhaustion.

The zinc(II) complex illustrated in Fig 13 is unambiguously monomeric as established from NMR spectra in solution. On the other hand a slightly less hindered analog illustrated in Fig 15 in which the geminal phenyl rings are not joined by a central bond is dimeric as revealed from the X-ray

Figure 15. Structure of a slightly less hindered tridentate ligand that clearly gives a dimeric complex (X-ray structure right).

structure. This illustrates vividly how great the tendency is for sulfide to act as bridging ligand. Unfortunately the complex shown in Fig. 13 is apparently sterically so well protected from oligomerization that potential substrate molecules and/or 1,4-dihydropyridines cannot approach it. Hence its catalytic activity is minimal.

The syntheses, not discussed here, of the tridendate ligands illustrated are challenging and time consuming. The systems remain very primitive compared to th order of complexity found in an enzyme like alcohol dehydrogenase. We are greatl encouraged, however, by the observation of intrinsic reactivity of these simple metal complexes. One cannot help wondering (and hoping) that these simple organometallic complexes model prebiotic complexes caught up into metabolism because of intrinsic reactivity and stability, and perfected through evolution.

The problem of translating enzyme structural information into chemical system that can be synthesized and tested under abiological conditions is great and in many instances still not surmountable. These cautionary words with regard to synthetic difficulties can be illustrated briefly with various dehydrogenase mode that have been developed and synthesized in recent years. One can scarcely overestimate the restrictions imposed on all this work by synthetic accessibility The summary we give here is in no way comprehensive; it is intended only to be illustrative and is supported largely by our own work.[25] The majority of effort has centered around derivatives of Hantzsch esters, prepared by condensation of tw beta-keto esters, an aldehyde (which becomes carbon-4 of the 1,4-dihydropyridine skeleton) and ammonia (see Fig. 16).

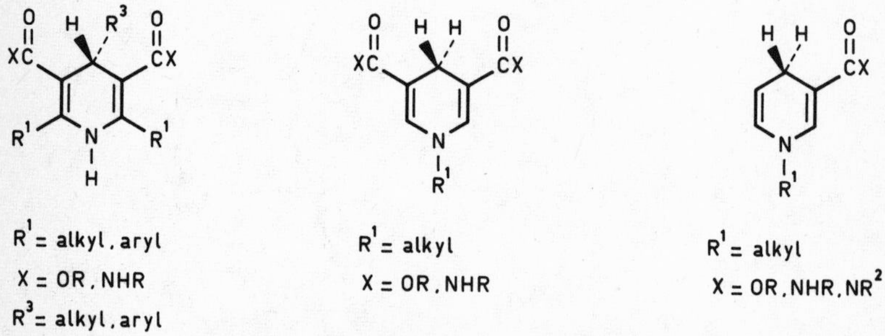

R^1 = alkyl, aryl
X = OR, NHR
R^3 = alkyl, aryl

R^1 = alkyl
X = OR, NHR

R^1 = alkyl
X = OR, NHR, NR^2

Figure 16. Commonly used 1,4-dihydropyridines. Left Hantzsch 1,4-dihydropyridine, middle derived from pyridine-3,5-dicarboxylic acid, right derived from pyridine-3-carboxylic acid.

Another system commonly employed is the 1,4-dihydropyridine derivable from commercially available pyridine-3,5-dicarboxylic acid. The third, and classical, 1,4-dihydropyridine is derived from pyridine 3-carboxylic acid. This is the basi heterocycle of nicotinamide.

The question of reactivity, already alluded to, has had a decisive influence
 the design of model systems designed to provide information about the working of
D(P)H cofactors. Virtually all of the 1,4-dihydropyridines derivable from the
arting skeletons shown in Fig. 16 have redox potentials near or exceeding that of
D(P)H itself.[26] However, the _kinetic_ reactivity (also for NAD(P)H) is decidedly
t great; usually only very electron deficient hydride acceptors are reduced
ontaneously. Activation is necessary. With the exception of the zinc and cobalt
gands described above, this has almost never been obtained in catalytic fashion
tside of the enzymic systems. In model reactions _activation_ can usually be
tained with the aid of _stoichiometric_ amounts of very potent electrophiles like
gnesium perchlorate in aprotic solvents. The basic idea is illustrated in Fig

gure 17

e function of the Mg(II) or other hard metal catalysts may be more subtle than
plied by this picture as pointed out by Ohno.[27] There is also precious little
reement as to exactly how the carbonyl group should be oriented relative to the
4-dihydropyridine during reduction. Judging from the many models that have been
veloped, and supported by theoretical calculations of Verhoeven,[28] a variety of
ientations for the carbonyl group with respect to 1,4-dihydropyridine may be
ceptable.

In general it has proved much easier in these synthetic systems to achieve
cognition rather than rate acceleration. This is only a generalization, of

course, and some exceptions can, and will, be found and discussed. Recognition generally entails recognition by a chiral 1,4-dihydropyridine-containing system o a prochiral face of a carbonyl compound. Seldom, however, is this recognition achieved concurrently with significant rate accelerations.

We have obtained at room temperature face selectivities of about 95% (90% enantiomeric excess) with the chiral bridged 1,4-dihydropyridines illustrated in Fig. 18 using alpha-keto esters as hydride acceptors.[25]

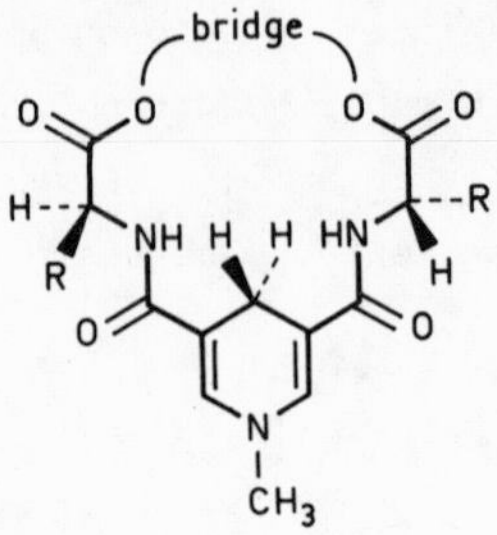

Figure 18

A stoichiometric amount of magnesium perchlorate is necessary as activator and there is no question of catalytic turnover of the model system. These systems provide the opportunity for a broad structure-reactivity investigation since the synthetic methods developed allow fairly ready change of bridging component and amino acid side chain. The original consideration in the development of these systems was complexation of the electrophilic activator in the macrocycle thereby allowing positioning of the carbonyl substrate in the manner illustrated in Fig. 19.

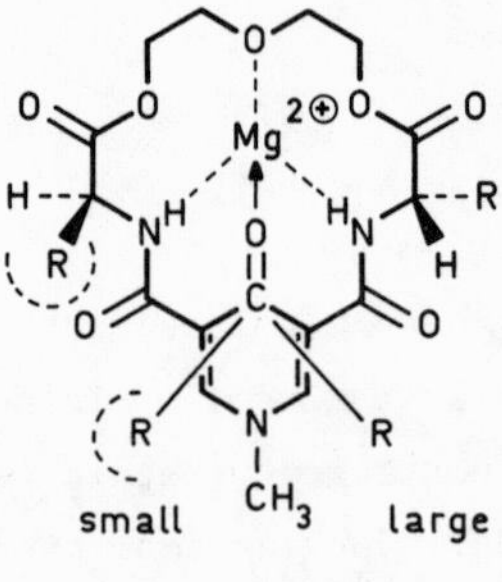

Figure 19

From this work it became clear that the complexing ability of the macrocycles is best modest and that the electrophilic activator is located more to the side than

ove the macrocyclic ring. Despite these modifications the stereochemical outcome
mains the same as originally postulated.

Restriction of conformational mobility of the macrocycle by choice of fairly
ort bridging segments holding the two lactone functionalities together leads, as
e anticipates, to higher degrees of recognition of prochiral faces of the
rbonyl component. An optimum is reached for 18- to 21-membered macrocyclic rings
ridge components of a length of 5-8 atoms). Also in accord with expectation the
ze of the group R on the amino acid has an effect on recognition; the larger the
oup the better the recognition.

Fortunately, there have recently appeared reports of model systems that
hibit high recognition together with, at least on qualitative grounds,
gnificant acceleration. The key is, not unsurprisingly, the use of
tramolecularity in the design of the systems. Meyers,[29] et. al., have used the
proach illustrated in Fig. 20. Enantiomeric excesses of 99% have been achieved
sing these modified 1,4-dihydronicotinamide derivatives generated intramolecularly
the presence of stoichiometric amounts of magnesium halide, which probably
erves as electrophilic activator.

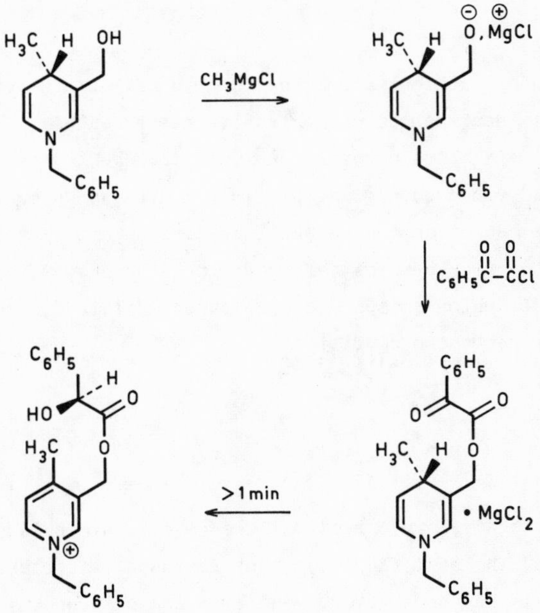

Figure 20

Another indication that significant rate accelerations can be achieved by th[e] device of intramolecularity is that of Kirby, et. al.,[30] who have observed spontaneous reduction in water-dioxane for the example shown in Fig. 21.

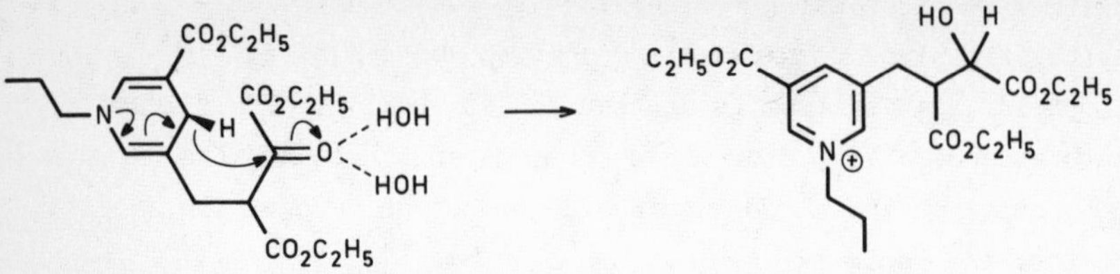

Figure 21. A model compound that reacts spontaneously in water-dioxane. The produ[ct] alcohol is, of course, racemic.

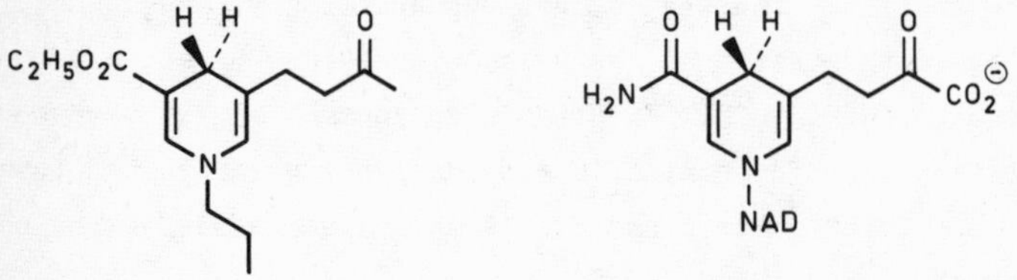

Figure 22

The compound is of course achiral and there is no question of recognition of prochiral faces. This Hantzsch ester derivative has a reactive carbonyl group placed in ideal, i.e. six atoms removed, proximity to the hydride at C-4. No met[al] catalysis for reduction is required; simply in 20% dioxane-water solution between pH 3-6 pH independent reduction occurs with a rate constant of about $2 . 10^{-3}$ sec^{-1}. Still the reactivity remains tempered; the other two 1,4-dihydropyridines illustrated in Fig. 21, which possess less electron deficient carbonyl groups, fai[l] to undergo reaction under these conditions.

C. Evolution, Models, and RNA?.

Some aspects of catalysis by nicotinamides have been considered so far and a very brief overview of the basic approach thus far usually followed for the design and synthesis of models has been given. The distance between these two approaches is considerable. Certainly we would like to unify somewhat better the two visions. We give here, again briefly, a framework on which this might be achievable. RNA assumes a large role in this vision.

Recent experimental work on template-directed oligonucleotide synthesis[31] and bioorganic chemical[32] and biochemical[33] considerations about the possible gnificance of ribonucleotide-like coenzymes and prosthetic groups for the origin

life, indicates the plausibility of a short but decisive evolution of

bonucleotide-built catalysts before the origin of the genetic code and

dependent of coded proteins.

We[34] have postulated that the phosphorylation necessary for the production of

ese RNA's could have occurred in vesicles. Plausible constituents of such

sicle membranes might be found among extant fat-soluble cofactors that still are

volved in bioenergetic processes like chlorophylls a, b, d etc., and

rresponding pheophytins, ubiquinones, plastoquinones, tocopherols, retinols and

rotenoids. All of these cofactors are fat-soluble by virtue of a polyprenyl

sed hydrophobic tail. Archaebacteria contain membranes based on similar

lyprenyl fatty molecules (diphytanyl glycerol ethers and dibiphytanyl-diglycerol

traethers).

Let us now turn to the specific case of nicotinamide. The (presumed)

talytically active conformation of nicotinamide is a structural analog of guanine

d could therefore form a base pair with cytosine as shown in Fig. 23. A model of

short RNA duplex in the left-handed Z conformation[35] has been built with the

gure 23. Guanine - cytosine base pair (left)
         Nicotinamine - cytosine base pair (right)

lp of Nicholson Molecular Models on a scale of 1 cm = 100 pm.[36] Total helix

lculations[37] show that the region of N-7 and O-6 of guanine (corresponding with

4 and carboxamide oxygen of nicotinamide (Fig. 23) has not only the highest

cessability, but also has the lowest molecular electrostatic potential, precisely

ere the positively charged hydrogen atom is situated in the transition state for

dride transfer. This means that a transition state like that in Fig. 11

ould be stabilized when nicotinamide is part of a Z double helix. Two stacking

se pairs, nicotinamide - cytosine above cytosine - guanine, are shown in Fig. 24.

e N-7 atom of guanine is a binding site for doubly charged metal ions.[35] The

sition of this metal ion relative to the C-4 atom of nicotinamide is very similar

the position of the zinc ion in alcohol dehydrogenase (Fig. 11).[38] A pyruvate

n, for example, can be fitted so that the metal ion bonds to the keto group of

pyruvate, and the carboxylate group is hydrogen bonded to the 4-amino group of
cytosine (Fig. 24).

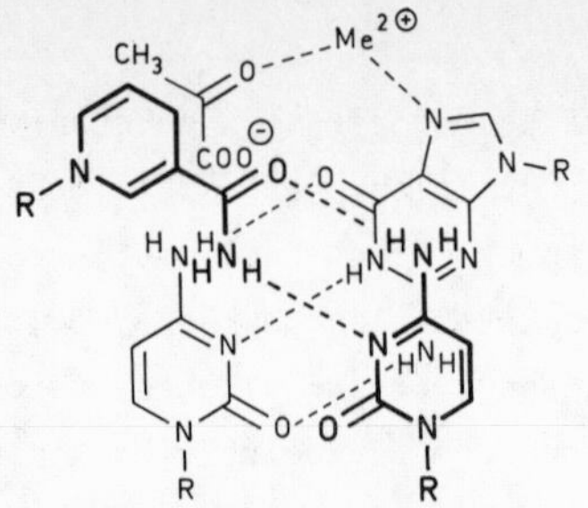

Figure 24. Two stacking base pairs of a RNA duplex in the Z conformation containi
nicotinamide as guanine analogue.

The relatively few nitrogen containing heterocycles that now function in DNA
and RNA were likely selected from a great number of possible heterocyclic systems
on the basis of their capacity for recognition. At the start of the evolution of
RNA molecules there probably were many more heterocycles that were part of RNA-li
molecules, among which flavins, deazaflavins, pterins etc. The relatively great
number of modified bases in t-RNA's are probably relics of that early period.

The structure of the Z helix is such that many of those molecules will fit i
double helical structures. For example, the extra substituted benzene ring in
flavin points outward away from the helix. The Z helix is only formed of molecul
with alternating purine-pyrimidine sequence. This is caused by the alternation
between syn and anti oriented bases. Pyrimidines do not easily take the syn form
For pyrimidine-like bases missing a 2-keto group, however, this is no longer the
case. This means that nicotinamide could also be found on places where normally
pyrimidine is present, when the nicotinamide is paired to, for instance,
pseudouridine (Fig. 25). There could exist an enormous number of potential

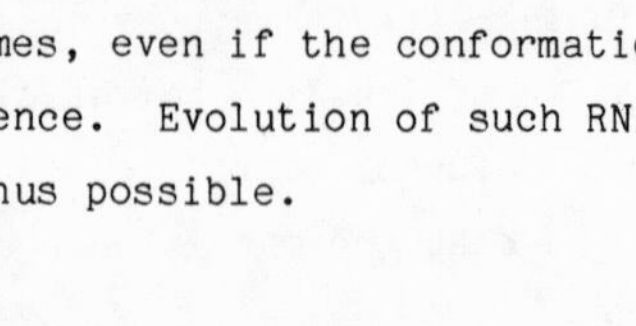

Figure 25. A nicotinamide-pseudo-uridine base pair.

prebiotic Z-RNA redox enzymes, even if the conformations of double helices were n
dependent on the base sequence. Evolution of such RNA redox enzymes on the basis
of point mutations seems thus possible.

The other possible conformation for double helical RNA, the A helix, has also
teresting properties with respect to catalysis of redox reactions. Again, total
lix calculations[38] show that the region of N-7 of adenine and guanine
orresponding to C-4 of nicotinamide) is at the same time the most accessible atom
 the base and borders upon the region of lowest molecular electrostatic
tential. Helices of the A type containing nicotinamide, flavin, etc., have thus
so to be considered as potential prebiotic redox enzymes. Nicotinamide can be
tted in this helix as a purine analog at all places, both as guanine and as
enine analog, depending on the conformation of the carboxamide group.

In Fig. 26 the base stacking of a flavin - cytosine base pair with a

gure 26. Stacking of successive flavin-cytosine and nicotinamide-cytosine base
         pairs in an A helix.

cotinamide - cytosine base pair is shown. The C-4 atom of nicotinamide lies
ove the C-4a atom of flavin. Similar stacking is observed in glutathione
ductase[39] and is expected to occur in all NAD(P)H-dependent flavoenzymes. A
milar enayme (ferredoxin : NADPH oxidoreductase) catalyzes the final step in the
otosynthetic NADPH production. Since flavins and deazaflavins are efficiently
otoreduced by almost everything that has ever been synthesized in abiogenic
stems, one could even speculate on the possibility of a primitive chlorophyll-
dependent RNA catalyst for photoreduction of $NADP^+$. The ferredoxin - NADPH
idoreductase of chloroplasts would then be the late descendent of those prebiotic
A catalysts.

Let us return now to some general considerations. Life on earth has probably
veloped to a great extent on the basis of a slow and continous process. The
ssibility of irregularities - jumps - is, of course, always present. In the slow
d continuous processes metabolic reactions have also been selected out of a large
ol of possibilities. The reactions must be suitable to be used in this
olutionary process and must meet therefore several conditions. First, it is
cessary that they can be catalyzed though interactions on the surface of peptide
ydrogen bridges, dipole-dipole interactions, prosthetic groups, etc.). Second,
e catalytic mechanism must be able to change in small steps ("improvement"). For

the most fundamental reactions (the oldest?), which from the very beginning of th
evolving (proto)metabolism have been present, these same two conditions apply als
in a much different context, namely with RNA as catalyst.The conditions become ev
more stringent because both the catalytic ability as well as the variability
(necessary for the "small steps") of RNA is much less than for peptides. Only
recently has serious thought gone into the consequences this must have had on the
development of reactions in living systems. Two excellent examples of the analys
that must be carried are those of Benner[41] and Bernhard[42] for the case of
dehydrogenases.

Translated into "testable chemistry" we expect that much can be gained by
designing and testing the reactivity of systems in which base pairing is used as
integral component of the structure. We know from enzymic chemistry that the
kinetic reactivity of 1,4-dihydronicotinamides, to restrict ourselves to a specif
example, can be unleashed by specific but as yet too poorly defined interactions.
It behooves us to define and use these interactions not only to understand enzymi
systems but to use enzymic principles more effectively in man-designed and man-
performed chemistry.

Footnotes and References

 1 a) Metzler, D.E., "Biochemistry", Academic Press, New York, N.Y.: 1977; b)
   Lehninger, A., "Biochemistry", Worth Publishers Inc., New York, N.Y.;1975.
 2 a) Klopman, G., J. Am. Chem. Soc, 1968, 90, 223; b) Salem, L. J. Am. Chem. Soc
   1968, 90, 543.
 3 a) Westheimer, F.H.; in Dolphin, D.; Poulson, R.; Avramovic, O. (eds), "Pyridi
   Nucleotide Coenzymes", Part A, Wiley, New York, N.Y.; p.p. 253-322; b) this
   series of two volumes provides one of the most succinct and up-to-date overvie
   of pyridine nucleotide chemistry and biochemistry now available. We will refer
   heavily to this source in this chapter.
 4 Verhoeven, J.W.; van Gerresheim, W.; Martens, F.M.; van der Kerk, S.M.;
   Tetrahedron, 1986, 42, 975.
 5 For a review, see Emsley, J.; Chem. Soc. Rev., 1980, 9, 91.
 6 a) Kirchen, R.P.; Ranganayakulu, K.; Rauk, A.; Singh, B.P.; Sorensen, T.S.; J.
   Am. Chem. Soc., 1981, 103, 588; b) Kirchen, R.P.; Sorensen, T.S.; Wagstaff, K.
   Walker, A.M.; Tetrahedron, 1986, 42, 1063.
 7 Jorgensen, W.L.; Salem, L.; "The Organic Chemist's Book of Orbitals", Academic
   Press, New York, N.Y.; 1973.
 8 Fleming, I.; "Frontier Orbitals and Organic Chemical Reactions", Wiley, London
   1976.
 9 Donkersloot, M.C.A.; Buck, H.M.; J. Am. Chem. Soc., 1981, 103, 6549.
10 See ref. 4 for a discussion.
11 Krechl. L.; Kuthan, J.; Coll. Czech. Chem. Commun.; 1981, 46, 740.
12 a) Grau, V.M.; in Everse, J.; Anderson, B.; You, K.-S. (eds), "The Pyridine
   Nucleotide Coenzymes, Academic Press, New York, N.Y., 1982, pp. 135-187.
13 See ref. 9 for a discussion.
14 De Kok, D.M.T.; Donkersloot, M.C.A.; van Lier, P.M.; Meulendijk, G.H.W.M.;
   Bastiaansen, L.A.M.; van Hooff, H.J.G.; Kanters, J.A.; Buck, H.M.; Tetrahedron
   1986, 42, 941.

a) One form of this equation gives the total interaction energy between molecules R and S as a sum of Coulomb effects, solution effects, and HOMO/LUMO interactions.

$$\Delta E_{total} = -q_r q_s \frac{T}{\varepsilon} + \Delta_{solv} + \sum_{m}^{occ} \sum_{n}^{unocc} \frac{2(c_r^m)^2 (c_s^m)^2 \beta^2}{E_m^* - E_n^*}$$

where q is the charge, T is a Coulomb repulsion term, ε local dielectric constant, β the overlap integral and $E_m - E_n$ the energy gap between the interacting HOMO/LUMO orbitals; b) another discussion is given in Dewar, M.J.S.; "The Molecular Orbital Theory of Organic Chemistry", McGraw-Hill Book Co., New York, N.Y., 1969.

For details see ref. 12.

These ideas, and other suggestions in this section, are developed in greater depth in Visser, C.M.; Life Science Advances D (Biochem.), accepted for publicatiaon.

Schirmer, R.H.; Schulz, G.E.; in ref. 3a, Part B, pp. 333-380.

Pai, E.F.; Schulz, G.E.; J. Biol. Chem., 1983, 258, 1752.

Hol, W.G.J.; van Duijnen, P.T.; Berendsen, H.J.C.; Nature (London), 1978, 273, 443.

a) Branden, C.-I., Jornvall, H.; Eklund, H.; Furugren, B.; in "The Enzymes", Vol. XI, 3rd edn., ed. Boyer, P.D.; Academic Press, New York, 1975, pp. 104-180; b) Eklund, H.; Nordstrom, B.; Zeppezauer, E.; Soderlund, G.; Ohlsson, I.; Boiwe, T.; Soderberg, B.-O.; Tapia, O.; Branden, C.-I., Akeson, A.; J. Mol. Biol., 1976, 102, 27.

a) Boersma, J. in "Comprehensive Organometallic Chemistry", ed. Wilkinson, G., Pergamon, Oxford, 1982, vol. 2, pp. 823-851; b) Holm, R.H.; O'Connor, M.J., Prog. Inorg. Chem., 1971, 14, 241.

The experimental approach is described in Kaptein, B.; Wang-Griffin, L.; Barf, G.; Kellogg, R.M.; J. Chem. Soc., Chem. Commun. 1987, 1457.

a) Zeppezauer, M.; Anderson, I.; Dietrich, H.; Gerber, M.; Maret, W.; Schneider, G.; Schneider-Bernlohr, H.; J. Mol. Catal., 1984, 23, 377; b) Sytkowski, A.J.; Vallee, B.L.; Proc. Natl. Acad. Sci. USA, 1976, 73, 344; c) Schneider, G.; Eklund, H.; Cedergen-Zeppezauer, E.; Zeppezauer, M.; Proc. Natl. Acad. Sci. USA, 1983, 80, 5289; d) Maret, W.; Anderson, I.; Dietrich, H.; Schneider-Bernlohr, H.; Einarsson, R.; Zeppezauer, M.; Eur. J. Biochem., 1979, 98, 501; e) Sartorius, C.; Gerber, M.; Zeppezauer, M.; Dunn, M.F.; Biochemistry, 1987, 26, 871; see also, for example; f) Makinen, M.W.; Maret, W.; Yim, M.B.; Proc. Natl. Acad. Sci. USA, 1983, 80, 2484; g) Makinen, M.W.; Yim, M.B.; ibid, 1981, 78, 6221.

Summarized in Talma, A.G.: Jouin, P.; de Vries, J.G.; Troostwijk, C.B.; Werumeus Buning, G.H.; Waninge, J.K.; Visscher, J.; Kellogg, R.M.; J. Am. Chem. Soc., 1985, 107, 3981 and the references contained therein.

Piepers, O.; Kellogg, R.M.; J. Chem. Soc., Chem. Commun., 1982, 403.

For example Ohnishi, Y.; Kagami, M.; Ohno, A.; J. Am. Chem. Soc., 1975, 97, 4766 and references summarized in ref. 25.

van der Kerk, S.M.; van Gerresheim, W.; Verhoeven, J.W.; Recl. Trav. Chim. Pays-Bas, 1984, 103, 143 and refs. 4 and 11.

a) Meyers, A.I.; Oppenlaender, T.; J. Am. Chem. Soc., 1986, 108, 1989; b) Meyers, A.I.; Brown, J.D.; J. Am. Chem. Soc., 1987, 109, 3155.

a) Kirby, A.J.; Walwyn, D.R.; Tetrahedron Lett., 1987, 28, 2421; b) references to the second structure shown in Fig. 22 are given in this article.

Inoue, T.; Orgel, L.E.; Science, 1983, 219, 859.

Visser, C.M.; Origins of Life, 1982, 12, 165.

White, H.B. III, in ref. 12, pp. 1-17.

Visser, C.M.; Origins of Life, 1984, 14, 291.

Wang, A.H.-J.; Quigley, G.J.; Kolpak, F.J.; van der Marel, G.; van Boom, J.H.; Rich, A.; Science, 1981, 211, 171.

36 For a discussion of details, see, Visser, C.M.; Origins of Life, 1984, 14, 301
37 Lavery, R.; Pullman, B.; Nucleic Acids Res. 1982, 10, 4383.
38 Eklund, H.; Branden, C.-I.; in "Dehydrogenases Requiring Nicotinamide
   Coenzymes", Birkhauser, Basel, 1980, pp. 41-84.
39 Corbin, S.; Lavery, R.; Pullman, B.; Biochim. Biophys. Acta, 1982, 698, 86.
40 Pai, E.F.; Schulz, G.E.; J. Biol. Chem., 1983, 258, 1752.
41 a) Benner, S.A.; Experientia, 1982, 38, 633; b) Benner, S.A.; Stackhouse, J.;
   "Chemical Approaches to Understanding Enzyme Catalysis", Green, B.S.; Ashany,
   Y.; Chipman, D.; (eds.), Elsevier, Amsterdam, Vol. 10, 1981, p. 32; c) Benner,
   S.A.; Nambiar, K.P.; Chambers, G.K.; J. Am. Chem. Soc., 1985, 107, 5513.
42 a) Srivastava, D.K.; Bernhard, S.A.; Biochemistry, 1984, 23, 4538; b)
   Srivastava, D.K.; Bernhard, S.A.; Biochemistry, 1985, 24, 623; c) Srivastava,
   D.K.; Bernhard, S.A.; Langridge, R.; McClarin, J.A.; Biochemistry, 1985, 24,
   629.

The Design of Peptides and Proteins
Ranging from Hormones to Enzymes

E.T. Kaiser

Laboratory of Bioorganic Chemistry and Biochemistry,
The Rockefeller University, 1230 York Avenue,
New York NY 10021-6399

Abstract: Our efforts on the modelling and redesign of proteins have
proceeded along two general routes. In one approach, the principal
subject of this article, we have focussed on designing and building
model peptides where we could neglect tertiary structure in at least
the early phases of our work. For many peptides and proteins that
bind in the amphiphilic environments of biological interfaces,
complementary amphiphilic secondary structures are induced. We have
developed design principles for the construction of model peptides
which have illuminated the roles of secondary structures in the
biological activity of apolipoproteins, peptide toxins and peptide
hormones.

On the other approach to protein engineering that we have pursued,
we have undertaken to redesign known tertiary structures by site-
directed mutagenesis or by chemical modification. In the latter
work we have introduced new covalently bound coenzyme analogs into
proteins, giving rise to semisynthetic enzymes with novel catalytic
activities.

Recently, we have embarked on a program of replacing secondary
structural units in folded proteins of known tertiary structure by
relatively non-homologous segments constructed employing design
principles similar to those used in our amphiphilic peptide work.
These studies represent a first step towards the design and construction of tertiary structure by the assembly of primary sequences.

S. A. Benner (Ed.)
Redesigning the Molecules of Life
© Springer-Verlag Berlin Heidelberg 1988

INTRODUCTION

The explosion of information over the last two decades relating structure to function in enzymatic catalysis is remarkable. Twenty five years ago, while reactive intermediates in a number of enzymatic reactions had been identified and functional groups involved in the catalytic act had been implicated by chemical modification experiments, the tertiary structures of enzymes were unknown. The determination of the structures of enzymes by x-ray crystallography, together with a variety of solution studies including, very importantly, kinetic measurements, have led to a situation where, at the present time, relatively detailed mechanisms which are at least reasonable can be proposed for most enzymes. Because of this progress and because of the revolutionary advances that have been made in the construction of biologically active molecules whether by mutagenesis procedures or by peptide synthesis, it has become feasible to embark on the redesign of structural and active site portions of enzymes, the catalytic molecules of life.

In my own laboratory until the late 1970's our research effort focussed heavily upon attempts to elucidate the nature of catalytic groups in reactive sites of enzymes and the mechanistic pathways by which the enzymes acted [1,2]. At that time, however, we decided that we should start to try to apply the lessons that we had learned from the study of naturally-occurring enzymes to the redesign of these catalytic proteins and ultimately to the design and construction de novo of enzymes. Of course, then as now, the major problem which confronted us in undertaking such research was the inadequacy of our ability to predict folding patterns (tertiary structure) from primary amino acid sequences [3]. Our recognition of this problem led us to embark upon two different approaches to the modelling and redesign of proteins. In one part of our effort, we concentrated on designing and building systems where we could neglect tertiary structure at least in the initial phases of our work [4,5]. While it is clear that even the ability to predict secondary structure from natural primary amino acid sequence using the various algorithms that are available is not highly reliable, it was our conviction that if we constructed relatively simple amino acid sequences employing those amino acids having the highest potential to form particular secondary structures, we could design such secondary structural units with some degree of confidence.

n our other approach, we have undertaken to redesign known tertiary structures by site-directed mutagenesis [6] (related work is being done in other laboratories) [7] or by chemical modification [8]. If the gene for an enzyme can be obtained fairly readily and if the protein can be expressed in a convenient system at high levels, then the site-directed mutagenesis methodology is very powerful and allows many structural variations to be made. A limitation of this methodology, however, is that substitutions of the natural sequence of amino acids must be made with other naturally-occurring amino acids. Also, when an amino acid substitution is made, there is always the possibility that a major change in tertiary structure may result when the nascent altered protein undergoes folding. The other methodology which we have employed which involves what we have called "chemical mutation" [8,9] is not particularly effective in the substitution of one amino acid by another in a structure except in selected cases (e.g., the conversion of the serine residue in subtilisin to a cysteine residue [10,11]) but it allows "post translational type modifications" to be made, such as the introduction of covalently bound coenzyme analogs which we have accomplished in our preparation of semisynthetic enzymes. Since the chemical mutation approach involves the modification of folded structures, generally, we do not expect gross changes to occur in the backbone structure of the proteins modified. However, some movement of residues in the vicinity of the modification site may be expected to occur, and, indeed, in preliminary x-ray crystallographic measurements on the structure of one semisynthetic species, flavohemoglobin, movement of part of the structure near the β-chain Cys-93 residue modified with the flavin has been found [12]. Furthermore, in our studies of flavoglyceraldehyde-3-phosphate dehydrogenase (flavo-GAPDH) we found that the quaternary structure of the enzyme underwent an apparent time-dependent change, presumably due to the steric bulk of the flavin molecule introduced on each subunit.

In most of our studies on the design of peptide models where we have focused on secondary structural considerations and in our site-directed mutagenesis studies, we have employed known techniques for the preparation of our peptide and protein systems. Recently, though, we have tested the hypothesis that the effective construction of a number of small proteins in which we are redesigning structural regions would be aided if we could prepare these proteins by stitching together the appropriate smaller peptide segments. Toward the end of this article we will describe methodology by which we can rapidly prepare side chain protected peptide fragment through the use of a solid phase support which we have developed and we will show that subsequent to the purification of thes segments they can be combined readily to produce small protein including enzymes and structural analogs thereof.

THE DESIGN AND CONSTRUCTION OF BIOLOGICALLY ACTIVE PEPTIDE MODELS BASED ON SECONDARY STRUCTURAL CONSIDERATIONS

Our decision to study models for biologically active peptides wher we could concentrate on the design of secondary structural units wa facilitated by the hypothesis that for many peptides and protein which bind in the amphiphilic environments of biological interface complementary amphiphilic secondary structures are induced [1,2] The molecule with which we initiated our modelling studies wa apolipoprotein A-I (apo A-I), the principal protein present in hig density lipoprotein (HDL) particles. It had been suggested tha this 243 amino acid containing protein which lies at the surface o the HDL particle contains structural regions which can be induced t form amphiphilic α-helices when binding occurs in the phospholipi milieu of the HDL particle [13]. Later, it was proposed that th amphiphilic α-helical regions are approximately 22 amino acids i length, that there are approximately 6 to 7 such regions in the ap A-I molecule and that the pattern of hydrophobic and hydrophili residues in the various helices is a repeating one [14,15]. Thi proposal seemed very appealing to us because one could construct reasonable picture for HDL in which the amphiphilic α-helica regions of apo A-I lay between the phospholipid head groups with th axes of the helices approximately tangentially to the surface an the hydrophobic faces of the helices penetrating somewhat into th particle in contact with the long chain hydrocarbon portion of th phospholipid molecules. We decided to test the structural proposa that amphiphilic helices were involved in the binding of apo A-I i HDL by constructing model peptides conforming to this hypothesis an by determining the biological and physical properties of thes peptides [16].

Our first objective was to see if we could make a new peptide whic would correspond to an idealized version of the putative amphiphili α-helical twenty-two amino acid region of apo A-I [17]. Therefore we undertook to construct the peptide using the following desig principles. First, we set out to have a hydrophobic/hydrophili balance in the model peptide which would correspond roughly to th elative ratio of these types of residues observed in the repeating elical pattern in apo A-I. Second, we tried to maintain the verall charge balance in the model helix as compared to that in the elical regions of the natural protein. Third, we used amino acid esidues with the highest α-helical potential in the respective ategories (hydrophilic positively charged, e.g., Lys, hydrophilic egatively charged, e.g., Glu, and hydrophobic aliphatic, e.g., eu). Fourth, and most importantly, we tried to reduce the sequence omology between the model peptide and the helical regions of the atural protein as much as possible. The reason for doing this is hat if we could show that an amino acid sequence very different rom that of the natural system had very similar physical and iological properties and the model sequence had been constructed on he basis of secondary structural design principles, the argument hat it was primarily secondary structure in the natural system hich was important rather than primary amino acid sequence in the egion in question would be strengthened enormously [19].

he first model we synthesized for apolipoprotein A-I was a 22 amino cid peptide, 1, which consisted principally of Leu, Lys, and Glu esidues with a C-terminal Ala residue, included for synthetic easons, and a N-Pro residue included because the helices in the atural system are punctuated at regular intervals by Pro and Gly. he design of the amphiphilic helix which we expected to be formed y 1 is shown in Figure 1 [17]. The physical properties of the odel undocosapeptide were found to resemble those of the total 243 mino acid apo A-I molecule. In particular, the model peptide 1 nderwent a concentration dependent aggregation giving a tetrameric pecies for which the apparent α-helicity was calculated to be imilar to that of apo A-I. Also, there was considerable similarity etween the behavior of the monolayers formed from the model and rom apo A-I at the air-water interface. The binding of the model eptide to egg lecithin vesicles exhibited saturation behavior iving a dissociation constant for complex formation that was pproximately only two-fold weaker from that seen for the interac-ion of the whole apo A-I molecule with vesicles. Most importantly, he activity in a biological system, the activation of the ecithin:cholesterol acyltransferase catalytic process was only omewhat less than that for the whole native apo A-I molecule. aken together, these observations indicate that an idealized amphi-hilic helical peptide structure with little sequence homology to he natural apo A-I sequence can exhibit the basic properties of the

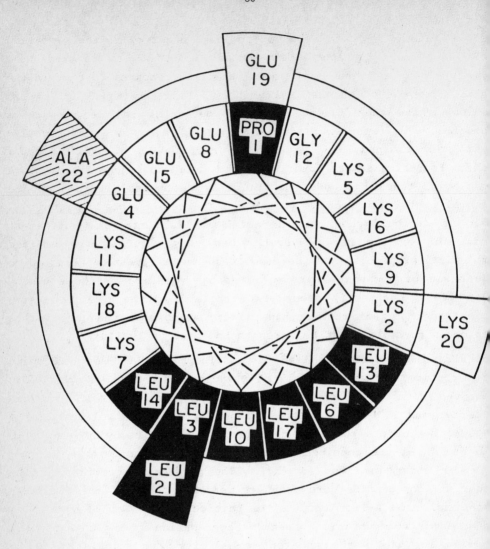

Figure 1. Axial projection of amphiphilic helical conformation of peptide 1. Hydrophobic residues are shaded.

hole molecule even though the model system is only 1/11 the natural
rotein's length.

hen binding was measured to cholesterol-containing vesicles in
hich the phospholipid/cholesterol ratio was 4:1, in contrast to apo
-I which actually binds better to the mixed vesicles, the model
eptide did not bind as well as it did to pure egg lecithin. When
e examined the proposed helical regions of apo A-I, we observed
hat, in some instances, there were arginine residues located in
hat otherwise would be hydrophobic regions of apo A-I. It seemed
ossible that the 3-OH group of cholesterol might have an unfavor-
ble interaction with the hydrophobic face of the amphiphilic
-helix that is inserted between the phospholipid head groups into
he vesicles. If this were to occur, the relatively favorable
nteraction of apo A-I with the mixed lecithin/cholesterol vesicles
ight depend on the presence of the polar arginine residues in the
redominately hydrophobic regions of the respective amphiphilic
-helices. To assess the role of the 3-OH function of cholesterol
n peptide-cholesterol interaction we incorporated an arginine
esidue into the hydrophobic region of the amphiphilic α-helical
eptide 1. As illustrated by Figure 2, peptide 2 contained an
rginine residue at position 10 in what would be otherwise a com-
letely hydrophobic region in the amphiphilic α-helix. We found
hat the binding of this peptide to cholesterol-containing vesicles
as significantly tighter than to pure egg lecithin vesicles [20].
hile our observations indicate that the difference in the binding
onstants for the interaction of peptide 2 with the cholesterol
ontaining vesicles and with the pure egg lecithin vesicles is
elatively small, the difference is still similar to the difference
een for apo A-I. Clearly, it is possible to fine tune the binding
f amphiphilic peptides to lipid or phospholipid surfaces through
he approach that we have developed.

n parallel with our design of model peptides to simulate the behav-
or of apo A-I we prepared and characterized peptides 22 and 44
mino acids in length based on putative amphiphilic helices in the
atural sequence of apo A-I [16,21]. We showed that a 44 peptide
egment containing the exact amino acid sequence corresponding to
esidues 121-164 of apo A-I mimicked the surface properties (i.e.,
ehavior as a monolayer at the air-water interface) of apo A-I more
losely than did the 22 peptide segment. Indeed, unlike the ide-
lized model 1, the natural sequence 22 peptide functioned very
oorly in mimicking apo A-I. The question arose, however, whether

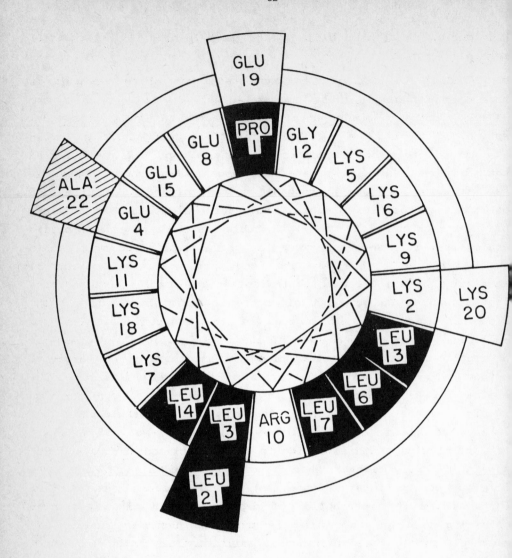

Figure 2. Axial projection of amphiphilic helical conformation of peptide 2. Hydrophobic residues are shaded.

he improved surface properties of the natural sequence 44 peptide
ere due merely to an increase in the size of the peptide or whether
he second 22 peptide segment present in the 44 peptide had an amino
cid sequence more conducive to reproducing the surface properties
f apo A-I. To explore this point we synthesized a 44 peptide model
(sequence shown in Figure 3), composed of two identical 22 peptide
egments corresponding to the sequence of peptide 1 [22]. We found
hat the covalent linkage of the two identical segments resulted in
considerable increase in the amphiphilicity of the peptide. Thus,
n 50% trifluoroethanol there is an increase in helicity for the 44
eptide as compared to the 22 peptide. Furthermore, there is a
onsiderably higher tendency of the 44 peptide to form aggregates,
uch as peptide micelles, in aqueous solution. Very important
onclusions have been drawn from measurements of the limiting
olecular areas of the respective peptides adsorbed at amphiphilic
nterfaces, at both the air-water interface as well as phospho-
ipid-coated polystyrene beads. Thus, we have found that the model
2 peptide occupies approximately 22 $\overset{o}{A}^2$ per amino acid at the
nterface. This indicates that the peptide is not fully helical and
ust contain some random coil segments, probably at the termini. In
ontrast, the 44 peptide occupies 14-16 $\overset{o}{A}^2$ per amino acid on both
urfaces in complete agreement with the value observed for apo A-I
tself. We have concluded that the true structural unit of the
polipoprotein is not really the 22 peptide segment but is actually
he 44 peptide segment which is punctuated in the middle by a helix
reaker either Gly or Pro [22]. Although this hypothesis remains to
e tested further, we have suggested that the Pro-containing α-heli-
es seen here and also for the strongly amphiphilic cytotoxic
eptide melittin are a new protein structural unit with physical and
iological properties distinct from a single α-helix.

ESIGN OF A PEPTIDE TOXIN

aving made appreciable inroads on the modelling of amphiphilic
econdary structures, we decided to design a peptide containing not
nly an amphiphilic secondary structural feature but also an active
enter. The candidate peptide which we decided to model was the bee
enom toxin melittin, a 26 amino acid peptide which is an activator
f phospholipase A_2 and which lyses erythrocytes. From the amino
cid sequence of the peptide and from information concerning binding
o phospholipids and other physical characteristics, we proposed
hat the NH_2-terminal 20 amino acids of the peptide might be forming

```
1                                           10
Pro-Lys-Leu-Glu-Glu-Leu-Lys-Glu-Lys-Leu-Lys-

                                  20        22
Glu-Leu-Leu-Glu-Lys-Leu-Lys-Glu-Lys-Leu-Ala-

23                                30
Pro-Lys-Leu-Glu-Glu-Leu-Lys-Glu-Lys-Leu-Lys-

                         40                 44
Glu-Leu-Leu-Gly-Lys-Leu-Lys-Glu-Lys-Leu-Ala
```

Figure 3. Amino acid sequence of peptide 3

quite hydrophobic amphiphilic α-helix [23]. One possibility was that the Pro residue present in this region might cause a kink in the helix. Alternatively, the proline might be flanked on either side by two shorter helical segments. We proposed that in addition to the amphiphilic ʌ-helix, there is a hexapeptide region at the COOH terminus containing a cluster of positive charge which acts as a sort of primitive active center. The truncated molecule melittin-(1-20) which lacks the hexapeptide portion, does not lyse erythrocytes but can bind to them.

In a test of our structural hypothesis we designed a new peptide, 4, illustrated in Figure 4, which had a sequence in which we reduced the homology very considerably relative to the native peptide in the H_2-terminal 20 amino acids [23]. Because we felt that the Pro residue probably was not a crucial factor in the lytic activity of the peptide, we decided to replace this residue with serine. On the hydrophobic face of the helix in model peptide 4 we employed Leu residues chosen for their high helix-forming potential, hydrophobicity, and electrical neutrality. While Gln would probably be the optimal choice for the neutral hydrophilic residues, we included some Ser residues to increase the hydrophilicity of the model. We retained the Trp residue at position 19 for studies of intrinsic fluorescence and we preserved the COOH-terminal hexapeptide active center portion of melittin.

From circular dichroism measurements on peptide 4 at neutral pH which indicated a marked concentration dependency of the mean residue ellipticity at 222 nm, we concluded that peptide 4 might undergo aggregation. Indeed, sedimentation equilibrium centrifugation showed that peptide 4 was tetrameric at a concentration of 2-3 $\times$ 10^{-5} M. Similarly, it is known that melittin forms tetramers. The circular dichroism data suggest that the helix content of peptide 4 is 69% for the tetramer and 35% for the monomer. For melittin itself, corresponding values are 48% and 18%. Stable monolayers at the air-water interface are formed by both peptide 4 and melittin. Surface pressure-area measurements showed discontinuities at 45 and 22 dyn/cm, respectively, presumably, the collapse pressures for the monolayers. Not surprisingly, the higher collapse pressure and the larger limiting area per molecule of peptide 4 compared to melittin indicated that peptide 4 is able to form a longer amphiphilic segment than melittin.

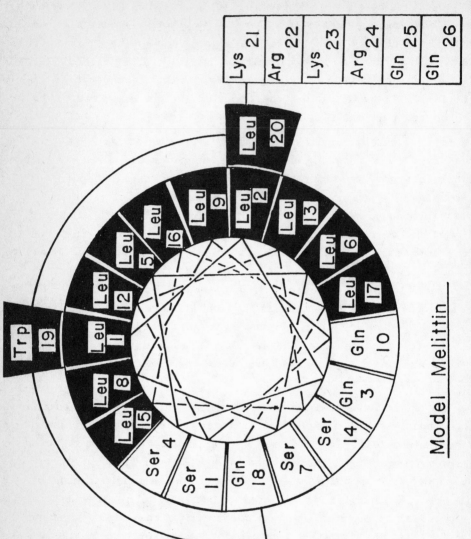

Figure 4. Axial projection of model melittin, peptide 4. Hydrophobic

t is interesting that the hemolytic activity of peptide 4 measured
n a 30 min incubation assay was significantly greater than that of
elittin. The surface affinity of peptide 4 is higher than that of
ative melittin, and this is consistent with the more extended
elical structure which we believe exists for the model peptide.
ndoubtedly, this is important for cell lysis. Our results indicate
oth our model and melittin caused a biphasic release of hemoglobin
rom erythrocytes at pH 7.3 and 4°C [24]. The model and melittin
roduced comparable fast phases on a molar basis, but the former was
ore effective in the slow phase. Employing a spectrophotometric
ethod involving the continuous monitoring of erythrocyte lysis, we
ound that when identical numbers of moles of melittin from stock
olutions of varying melittin concentrations were added to an ery-
hrocyte suspension, the rate of hemoglobin release showed a bell-
haped dependence on the concentration of the stock solution [25].
he maximum rate of lysis was found near the concentration at which
elittin changes from the predominately monomeric to the predomi-
ately tetrameric state. From an analysis of the concentration
ependence, it is clear that melittin undergoes a reversible cooper-
tive tetramerization, and it appears that the hemolytically active
pecies is the dimeric form. Our observations support the hypothe-
is that the amphiphilic helical form of melittin is the species
ith membrane affinity, because the monomer is largely in the random
oil form, while in the tetrameric aggregate the micellar structure
ould prevent contact between the membrane and the hydrophobic
ortion of the melittin helix. It should be mentioned that the
etrameric form of melittin has been crystallized and its structure
etermined in two crystal forms to 2.0 and 2.5 Å resolution [26-28].
he x-ray data for the crystals were consistent with the structural
ypothesis we had proposed earlier in designing the melittin model
escribed in this section of this review. In both of the crystal
orms, a bent amphiphilic helical rod structure is exhibited by
elittin. Hydrophobic interaction of the apolar side chains is
bserved in the aggregated form of the four bent amphiphilic heli-
es.

ur results indicate that the functional units sufficient for the
ctivity of melittin-like cytotoxic peptides are a 20 amino acid
mphiphilic α-helix with a predominantly hydrophobic face and a
hort segment with a high concentration of positive charges. In
his context, it is interesting to note the δ-hemolysin, a cytotoxic
eptide from Staphylococcus aureus is very similar in its properties

to melittin. It forms aggregates with highly α-helical characte
and also produces very stable monolayers at the air-water interfac
[29]. Furthermore, it interacts strongly with phospholipids [30
and serum lipoproteins [31]. Like melittin, δ-hemolysin appears t
involve in its active conformation an N-terminal amphiphilic helica
region and a C-terminal cluster of positively charged residues.

DESIGN AND CONSTRUCTION OF PEPTIDE HORMONES BASED ON SECONDARY
STRUCTURAL CONSIDERATIONS

We have proposed that biologically active peptide hormones can b
placed in three general structural categories [5]. In the firs
category are short peptides such as [Met^5]-enkephalin, [Leu^5]-en
kephalin, and thyrotropin-releasing hormone. The "active sites" o
specific recognition sites that determine the interaction of thes
hormones with their receptors comprise essentially their whol
structures. In the second category of peptide hormones, we hav
included structures of greater complexity that are large enough t
be stabilized in aqueous solution in their three dimensional form
through crosslinking by means of multiple disulfide bonds or b
formation of a suitable hydrophobic core. Among these peptides ar
insulin and growth hormone. In the third category are peptides tha
typically consist of a single chain of about 10-50 amino acids an
that generally do not contain extensive disulfide crosslinkin
(usually not more than one disulfide linkage). When these peptide
act at biological interfaces, the environment that they encounter i
often amphiphilic. For peptide hormones belonging to the thir
structural category, we have proposed that the binding of thes
molecules to the interfaces is likely to give rise to the inductio
of complementary amphiphilic secondary structural regions [23,32].

Several important roles for amphiphilic secondary structures i
peptide hormones seem likely [33,34]. For example, such structure
may function to position other parts of the hormone, like th
specific recognition site, in an orientation resulting in productiv
interactions with the receptor, giving rise to signal transmission
Another possible role for the amphiphilic secondary structure may b
in guiding the peptide to its receptor quickly. In aqueous solutio
the peptide hormones of the type we are discussing typically do no
have much structure. Once they encounter a biological surface i
the vicinity of their respective receptors, the secondary structur
of the peptide hormone may be induced, allowing the peptide to fin
its receptor by adsorption to the cell-surface, followed by diffu

ion in only two dimensions on that surface. This could result in a onsiderable increase in the rate at which the hormone finds the eceptor as compared to what would be involved in a three dimension- l search. Another point which must be mentioned is that the mphiphilic secondary structure can play a role in protecting the eptide hormone from proteolysis. It seemed possible that for at east some of the peptide hormones intramolecular folding of the tructure may occur. Such folding would allow the amphiphilic econdary structure to interact with other portions of the molecule, his intramolecular interaction which would give rise to a quite rdered structure would undoubtedly provide protection from proteo- ytic attack. Also, if the amphiphilic secondary structural region nteracts with the membrane surface strongly, this could also help o stabilize the hormone against enzymatic attack.

n the basis of structural modelling using space filling models as ell as computer graphics, possible amphiphilic helical regions have een identified for numerous peptide hormones in the third struc- ural category mentioned above. Among such peptide hormones are alcitonin [35-37], calcitonin gene-related peptide [38], cortico- ropin releasing factor [39], β-endorphin [32,34,40-44], glucagon 45,46], growth hormone releasing factor [47], neuropeptide Y [34], ancreatic polypeptide [34], parathyroid hormone [34,48], secretin, nd vasoactive intestinal peptide [49]. While these peptides have n common the possible formation of amphiphilic helical regions, hey are very different in many structural characteristics. Thus, heir hydrophobic domains in the amphiphilic helical conformation ary in size and shape. Furthermore, the numbers and kinds of harged residues on the hydrophilic face of the helix, the way in hich the aromatic residues are placed on the hydrophobic face, the ength of the helix and the number of "mistakes" in the location of esidues (e.g., hydrophilic residues in the hydrophobic domain) vary rom case to case. Despite the frequent occurrence of the amphi- hilic helical structures in the peptide hormones, regions of poten- ial amphiphilic β-strand structure occur relatively seldomly [34]. enerally speaking, when they do occur such regions involve se- uences which are 10 or fewer amino acids in length. They appear to nclude regions of important hormones such as luteinizing hormone eleasing hormone and dynorphin A (1-17). We have suggested [34] hat amphiphilic β-strands may not be encountered as frequently in eptide hormones as our amphiphilic helices, at least in part, ecause the β-strands tend to aggregate very strongly and they can

be essentially intractable in peptide systems when stretches of mor
than a few amino acids are involved.

In our work we have developed models in which the putative secondar
structural segments of peptide hormones have been replaced wit
relatively non-homologous amino acid sequences which preserve th
most important features of the structural regions, usually in a
idealized form. We have developed structure-function relationship
which establish the importance of each structural feature throug
the comparison of the physicochemical and pharmacological propertie
of these models with those of the corresponding natural hormones
If we can show that homology with the natural sequence of a particu
lar structural element can be successfully minimized and the proper
ties of the peptide dependent on that element are still reproduce
in a satisfactory manner, this provides strong evidence that w
have, indeed, proposed the correct biologically active conformatio
of the hormone. A few examples in which we have probed peptid
hormone conformation by the design of suitable models are no
discussed.

One of the most striking examples of the importance of an amphiphil
ic α-helical structure can be found in the case of the peptid
hormone calcitonin, which has hypocalcemic activity [35,36,50]
From literature data on analogs and on the basis of the inspectio
of molecular models, we have proposed that the biologically activ
structure of calcitonin may consist of three regions: an "activ
site" or "specific recognition site", the amino terminal heptapep
tide region which has a disulfide loop between Cys-1 and Cys-7; a
amphiphilic α-helical domain containing residues 8-22; and a hydro
philic carboxyl-terminal region containing residues 23-32. I
stringent tests of this proposed model, we designed and charac
terized three 32 amino acid peptide models for calcitonin, MCT-I
MCT-II and MCT-III. As shown in Figure 5 which gives an axia
projection of the helical region between residues 8-22 ir salmo
calcitonin I, the most active form of the hormone, there is
remarkable segregation of residues with a clustering of hydrophilic
on one face of the helix and hydrophobics on the other. The onl
residue which is "out of place" is the Glu at position 15 whic
interrupts an otherwise completely hydrophobic face. Indeed, th
grouping of Leu residues on the hydrophobic face is striking sinc
among the naturally-occurring aliphatic amino acids, this residue i
the best α-helix former. In much of our work on the design o
models for peptide hormones, we have utilized naturally-occurrin

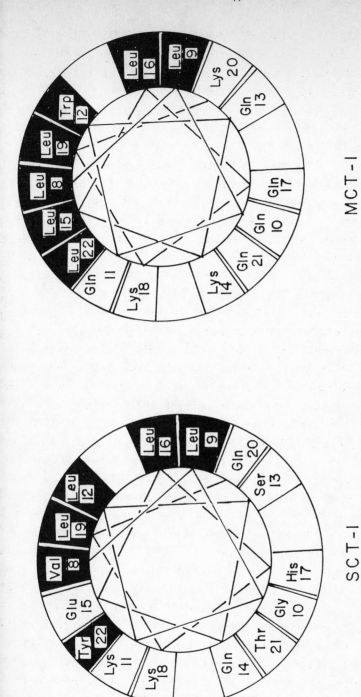

Figure 5. Axial projections of amphiphilic helical conformations of salmon calcitonin I (SCT-I) and model calcitonin (MCT-I). Hydrophobic residues are shaded.

amino acids. While the solid phase peptide synthesis methodology w
have employed in the construction of our peptides can be used t
introduce unnatural amino acids, we focused on natural residues fo
the most part because we wanted to develop approaches which coul
then be applied to systems where we would use genetic engineerin
methodology to construct our models. Because the hydrophobic fac
of salmon calcitonin is already close to what would be expected t
be an idealized structure, we don't have a great deal of flexibilit
in altering the sequence in this region. Nevertheless, in our firs
model of MCT-I [35] we prepared a system where not only did we alte
the hydrophilic face of the 8-22 sequence, but also we tried t
change the hydrophobic face appreciably (Figure 5). The changes w
made on the hydrophobic face of the first model included a substitu
tion of Trp for Leu-12, a change of the Tyr-22 residue to Leu, an
replacement of the Glu-15 residue by a Leu. Additionally, a smal
change was made when the Val-8 residue of the natural hormone wa
replaced by a Leu residue. Measurements on the biological activit
of MCT-I showed that this model does bind specifically to calcitoni
receptors in rat brain particulate fractions, and rat kidney corti
cal membranes, although it is somewhat less effective than salmo
calcitonin. The model does have potent hypocalcemic activit
roughly comparable to that of porcine calcitonin, as monitored in
rat bioassay. We have concluded that our results make a strong cas
that the region from residues 8-22 of calcitonin has primarily
structural role as an amphiphilic α-helix interacting with th
amphiphilic environment of the calcitonin receptor. In peptid
model MCT-II (Figure 6) we utilized the same hydrophilic amino aci
distribution as in MCT-I, but we maintained more of the hydrophobi
residues present in salmon calcitonin [36]. In particular, w
retained residues Leu-12 and Tyr-22 of the salmon hormone. We four
that both the binding of MCT-II to calcitonin receptors and it
hypocalcemic activity were very reminiscent of the behavior of th
salmon hormone, even though the hydrophilic face of the amphiphili
helical region of the model differed significantly from that of th
naturally-occurring peptide. The structural hypothesis we have mad
for calcitonin makes it possible to proceed very systematically t
test the importance of particular structural features. In our mos
recent model, MCT-III (Figure 6), we put back in the Glu-15 residu
in the otherwise hydrophobic face [50]. In other words, in mode
MCT-III it is really only the hydrophilic face of the 8-22 regio
which has been idealized and in which the sequence homology to th

A

SCT-I   H$_2$N-CYS-SER-ASN-LEU-SER-THR-CYS-VAL-LEU-GLY-LYS-LEU-SER-GLN-GLU-LEU-

MCT-II   H$_2$N-CYS-SER-ASN-LEU-SER-THR-CYS-LEU-LEU-GLN-LEU-GLN-LYS-LEU-LEU-

MCT-III   H$_2$N-CYS-SER-ASN-LEU-SER-THR-CYS-LEU-LEU-GLN-GLN-LYS-GLU-LEU-

HIS-LYS-LEU-GLN-THR-TYR-PRO-ARG-THR-ASN-THR-GLY-SER-GLY-THR-PRO-NH$_2$

GLN-LYS-LEU-LYS-GLN-TYR-PRO-ARG-THR-ASN-THR-GLY-SER-GLY-THR-PRO-NH$_2$

GLN-LYS-LEU-LYS-GLN-TYR-PRO-ARG-THR-ASN-THR-GLY-SER-GLY-THR-PRO-NH$_2$

B

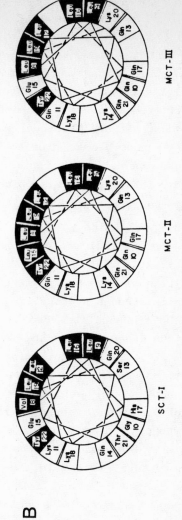

Figure 6A. Sequences of SCT-I, MCT-II and MCT-III.

       B. Axial projections of amphiphilic helical regions
          of SCT-I, MCT-II and MCT-III.

natural hormone has been drastically reduced. We have found wit
this latest model that the hypocalcemic activity is actually some
what elevated (about 2-1/2 to 3-fold) relative to the salmon hor
mone, although it was less effective than the naturally-occurrir
hormone in receptor binding. Presumably, this difference in th
hypocalcemic activity and in the observed receptor binding potenc
may reflect differences in tissue selectivity for the salmon hormon
and for the model. In any case, our work with the calcitonin model
shows that it is possible to proceed systematically in the design
construction and evaluation of peptide hormone models utilizing th
amphiphilic structural hypothesis which we have proposed.

The hormone which we have studied most extensively in our laborator
is β-endorphin, a 31 amino acid peptide, which has opioid activit
[32,34,40-44]. Accordingly to our structural hypothesis, there ar
three important regions in β-endorphin. First, there is a specifi
opioid recognition site comprising residues 1-5 which corresponds t
[Met[5]]-enkephalin; then a hydrophilic spacer covering residues 6-12
and finally, the sequence from Pro-13 to residue 29 which is almos
at the C-terminus and which has the potential to form an amphiphili
helix. Unlike the case of calcitonin, there is some ambiguity as t
precisely which type of helix the C-terminal region forms in β-er
dorphin. In particular, besides the α-helix, it seems conceivabl
that an amphiphilic π-helix could be formed. The difference betwee
these two structural proposals is illustrated in Figure 7 whic
shows helical net diagrams. In the α-helical structure there ar
3.6 residues per turn whereas in the π-helical structure there ar
4.4. Pictured in the α-helical form, the hydrophobic domain of th
region comprising residues 13-29 would curve along the length of th
helix. However, in the π-helical form the hydrophobic domain woul
lie straight along the length of the helix axis, just as is observe
with the α-helix for calcitonin. Generally speaking, x-ray struc
tural data on globular proteins do not show the presence of π-heli
cal structure. However, it seems to us that it is still possibl
that amphiphilic π-helical structures could be formed on a surfac
in an amphiphilic environment.

Since little is known about π-helical structures, it is difficult
predict which residues would be most likely to go into a π-helic
conformation in a model peptide system. For this reason we hav
chosen to model the amphiphilic α-helical structures in our work c
the design of β-endorphin analogs. We have designed six peptic
models of β-endorphin for which α-helical net diagrams of the 13-

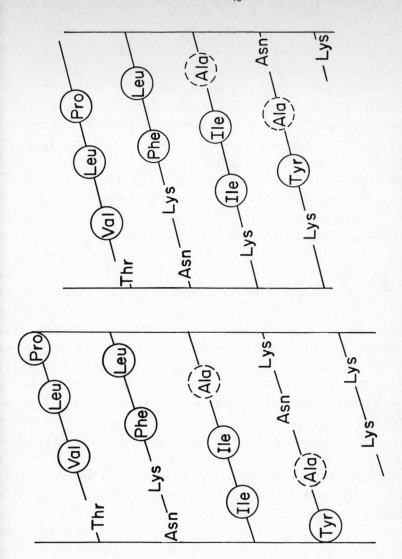

Figure 7. Helical net diagrams of region including residues 13–29 in β-endorphin. Circled residues are hydrophobic groups. Left-hand drawing: α-helix. Right-hand drawing: π-helix.

region are shown in Figure 8 [34]. In peptide 5 the helical regio
from residue 20-31 is idealized and is comprised of only Gln, Ly
and Leu residues. In peptide 6 the whole helical region is ide
alized. For both peptides 5 and 6 the shape of the hydrophobic fac
has been changed from that in β-endorphin since in the two mode
peptides the hydrophobic face lies essentially along the helix ax:
rather than curving as appears to be the case in the α-helical fo?
of β-endorphin. The behavior of peptides 5 and 6 in binding to bo?
the δ- and μ-opiate receptors is roughly comparable to that o
β-endorphin, and this behavior is consistent with the hypothes:
that the C-terminal segment of β-endorphin does not play a determin
ing role in the binding of these types of opiate receptors [32,40]
Similarly, it was found that the inhibitory effects of peptides
and 6 on the contractions of electrically stimulated guinea pi
ileum were similar to those of β-endorphin. It is known that th
C-terminal region of β-endorphin is important for the inhibition o
the twitching of electrically stimulated rat vas deferens sino
truncated forms of the hormone lose activity. Therefore, ou
findings that both peptide 5 and 6 were effective inhibitors in th
rat vas deferens systems were important in demonstrating that a
idealized amphiphilic helical structure with little homology to th
natural sequence was functional in this assay. As in the case o
β-endorphin, the inhibitory effects of peptides 5 and 6 could b
reversed by the addition of the opiate antagonist naloxone. At th?
point the very similar pharmacological behavior of peptides 5 and
with that of β-endorphin concerned us somewhat. Clearly, the shap
of the hydrophobic face of the helical region in these peptides wa
quite different from that in β-endorphin, and this raised the issu
whether the nature of the amphiphilic helix in the C-terminal regio
of β-endorphin was really important in determining its physiologica
properties. To address this issue, we next prepared peptide model
in which we introduced a curved hydrophobic face in the idealize
α-helical section [41]. As with models 5 and 6, the behaviors o
peptide 7 in binding to μ and δ receptors and inhibiting the con
tractions of either guinea pig ileum or rat vas deferens wer
reminiscent of the behavior of β-endorphin. However, an importar
feature of β-endorphin which peptides 5 and 6 did not mimic was th
analgesic activity. Neither peptide 5 nor peptide 6 was an analges
ic agent, whereas peptide 7 showed a marked antinociceptive effec
in mice which could be reversed by naloxone. The analgesic potenc
of peptide 7 is somewhat less than that of β-endorphin and the onse

47

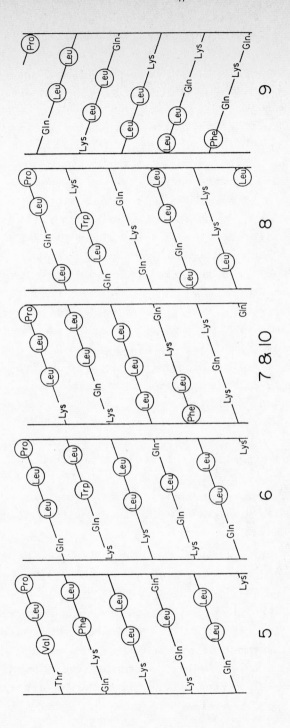

<u>Figure 8.</u> Helical net diagrams of α-helical representations of the 13-31 region in peptide models of β-endorphin. Circled residues are hydrophobic.

of the maximal effect of peptide 7 was slower than what was observe
for β-endorphin, but the model peptide showed a considerably longe
lasting effect. It appears, thus, that proceeding systematically i
examining not only the idealization of the helical region but als
the shape of the hydrophobic face, one can discern which among th
properties of β-endorphin depends primarily on the N-terminal regio
of the hormone, which of them requires an amphiphilic helix in th
C-terminal region and which requires a curved hydrophobic face i
the latter region. In our work on β-endorphin analogs, we have als
constructed a "negative" model 8. It was designed in order t
determine how the lack of a well-defined amphiphilic structur
affects the biological properties of β-endorphin [42]. This peptid
contains the three structural units previously postulated for β-en
dorphin, but in the 13-31 region the amino acids are arranged in
way such that no definite continuous hydrophobic zone could b
formed in either an α- or a π-helical conformation of this region
This model peptide was able to reproduce a number of the propertie
of β-endorphin including strong binding to both δ or μ opiat
receptors, and high activity in opiate assays on guinea pig ileum
In contrast, in assays on the rat vas deferens which are ver
specific for β-endorphin, the potency of the model peptide was ver
low, and it could be shown that it was not mediated by the sam
opiate mechanism or by the same opiate receptor. Thus, our result
indicate that the amphiphilicity of the helical structure in residues 13-31 is essential for high opiate activity on the rat va
deferens and the receptors therein, whereas no such structura
requirement appears to be necessary for interaction with the opiat
receptors on a guinea pig ileum.

In a further model, we constructed a peptide in which the natura
sequence of β-endorphin was retained in residues 1-12, but D-amin
acids were used in the C-terminal segment 13-31 [43]. Our previous
ly studied peptide 7 was used as a basis for the construction of th
hydrophobic domain of peptide 9. If a left-handed α-helical confor
mation for the 13-31 segment of peptide 9 is assumed, the respectiv
positions of the D-amino acids in the sequence have been chosen s
that they allow the formation of an amphiphilic α-helix with
hydrophobic domain quite similar in size and shape to that o
peptide 7. It is interesting to note that, in the rat vas deferen
assay which we have already mentioned is very specific for β-endor
phin, peptide 9 shows a mixed agonist-antagonist activity. Peptide
9 shows a potent analgesic effect when injected intracerebroven-

ricularly into mice. Although at equal doses the analgesic effect
f peptide 9 is only about 10-15% of that of β-endorphin, it is
onger lasting. Taken together with our earlier model studies, our
esults on peptide 9 show clearly that the amphiphilic helical
tructure in the C-terminus of β-endorphin is of predominant impor-
ance with regard to activity in rat vas deferens and analgesic
ssays. The observed similarity in the in vitro and in vivo opiate
ctivities of β-endorphin and peptide 9, despite the drastic change
n chirality in the latter model, demonstrates that even a left-
anded amphiphilic helix can function satisfactorily as a structural
nit in a β-endorphin-like peptide.

he most recent model we have built involves the use of unnatural
uilding blocks [44]. In this peptide, 10, the hydrophilic linker
egion between the NH_2-terminal enkephalin (residues 1-5) and the
arboxyl terminal helix (residues 10-28, sequence identical to that
f peptide 9 in region 13-31) consists of 4 units of γ-amino-γ-
ydroxymethylbutyric acid connected by isopeptidic linkages. The
roperties of peptide 10 in a monolayer at the air-water interface
re similar to those of peptide 9, as is this peptide's circular
ichroism behavior. The binding affinities of peptides 7 and 10 are
lso similar. In rat vas deferens assays peptide 10 was equipotent
o peptide 7. Most importantly, peptide 10 shows potent analgesic
ctivity when injected intracerebroventricularly into mice. Our
indings with peptide 10 clearly demonstrate that unusual building
locks can be utilized in the construction of structural regions of
ynthetic analogs with the preservation of the biological activity
f peptide hormones.

his is not meant to be an exhaustive description of our study of
mphiphilic peptide hormones, and only two more examples will be
riefly described. In amphiphilic environments such as the
ir-water interface and the surfaces of phospholipid vesicles,
orticotropin-releasing factor (CRF) assumes an amphiphilic secon-
ary structure [39]. The construction of molecular models suggests
hat the CRF molecule consists of two regions of high helical
otential in which the hydrophilic and hydrophobic residues are
egregated on opposite sides of the cylindrical helix. From our
odel building and from our results on the binding of CRF to model
mphiphilic environments, we have proposed that it is the helical
orm that binds in the biologically active conformation to cell
embranes.

In the case of the human growth hormone-releasing factor (GHRF), i
is not entirely clear how many residues at the N-terminal region ar
included in the active site. In our building of models we have bee
rather conservative and have not varied the first seven residues
although it is likely that the active site region of the hormone i
actually no more than four or five residues long. We have suggeste
that the region from residue 7-29 of GHRF has the potential to for
either an amphiphilic α-helix where the hydrophobic face twist
along the helix axis or an amphiphilic π-helix where the hydrophobi
face lies straight along the axis [47]. The hormone has typicall
amphiphilic properties, forming a very stable monolayer at th
air-water interface and binding tightly to unilamellar vesicles. W
have built a 29 amino acid GHRF analog where there has been
maximization of the potential of the 7-29 region to form an amphi
philic α-helix. Just as in the proposed α-helical structure of thi
region in the natural hormone, the hydrophobic face twists along th
helix axis in the analog structure. In our attempt to reduc
sequence homology between the natural sequence and the model, 1
residues in the 7-29 region have been altered. We have shown tha
the C-terminal amidated 29 amino acid analog was 1.6 times as poter
as GHRF (1-40) OH and the free acid form of the analog was 1/6 a
potent in an assay of growth hormone secretion stimulation i
primary cultures of rat anterior pituitary cells. Our result
provide strong support for the hypothesis that there is an amphi
philic helical region in the C-terminal portion of biologicall
active GHRF. It should be pointed out here, however, that since w
don't know how to construct π-helical structures with confidence
the findings we have made in the cases of both β-endorphin and GHF
which could in principle form either α-helices or π-helices th
question has been left open whether there is, indeed, a π-helica
contribution to the biologically active conformation.

TOWARD THE DESIGN AND CONSTRUCTION OF SMALL PROTEINS INCLUDING
ENZYMES

In this section of the article we will cover work which is far frc
complete. There are two questions which we wish to address here
First, having shown that we can design secondary structural units i
many systems which show interesting biological properties, we woul
like to approach the problem of designing tertiary structure. A
discussed earlier, our ability to predict tertiary structure frc
primary amino acid sequence is not well developed. For this reasor

we have focused on an empirical approach which gives us at least the beginnings of the ability to design tertiary structure. Specifically, we have used as a starting point folded structures which exist in nature and have raised the issue whether or not it is possible to replace secondary structural units in such folded proteins by relatively non-homologous segments constructed with design principles similar to those which we have employed in our amphiphilic peptide work. The other question raises the issue of the means by which one might construct small proteins. There are a number of possible answers. One would be to use molecular biological techniques. If the gene for the protein can be cloned and the protein can be expressed at high levels, this is a rather attractive route. Alternatively, one can pursue the synthesis of the gene for the protein and then try to express the protein in a suitable system such as E. coli or yeast. Of course, there are a number of possible difficulties with either of these approaches. For example, there are many cases of proteins which cannot be expressed readily in a convenient system. Indeed, one of the enzymes with which we are currently carrying our redesign studies, ribonuclease T_1, provides an example of the limitation, at least with current technology, in the genetic approach [51]. In this instance the genes for the natural enzyme as well as for several mutants, have been synthesized, but the expression of the mutant enzymes in E. coli does not proceed very well because the enzymes are deleterious to the bacteria. Another limitation in the use of the genetic engineering methodology is that this technology is only worked out at the present time for the 20 naturally-occurring amino acids. It is not feasible to introduce unnatural amino acids by site-directed mutagenesis. Yet, a further difficulty is that if one wishes to introduce suitable labels for spectroscopic studies, this is difficult to do selectively with the genetic engineering methodology. In other words, if one wished to introduce a ^{13}C label in a particular amino acid residue in a protein, one would end up typically labelling all residues corresponding to that amino acid.

In principle, it might be considered feasible to prepare small proteins on the order of 100 amino acids in length through stepwise solid-phase synthesis. The preparation of proteins of this length has been reported [52]. However, there are great difficulties with such syntheses. First, in the stepwise synthesis of a protein of this length it is inevitable that many impurities will be present at the end of the synthesis. Not only will purification often be

difficult, but also establishing the purity of a protein synthesize
in this way may not be something which can be readily done. Addi
tionally, if one wishes to redesign significant portions of
protein molecule, then each time a new analog is made it may b
necessary to proceed through, effectively, a total synthesis of th
mutant molecule. For these reasons, we have been trying to develo
methodology which will allow us to prepare small proteins includin
enzymes with considerable flexibility in the introduction of altere
structures. The approach we have taken is to focus on the rapi
synthesis of protected peptide segments usually not more than te
amino acids in length, followed by their purification and then th
assembly of these segments into larger segments which are the
finally coupled to give the entire protein [53]. This genera
approach has a great deal of flexibility. Because small segment
are put together to make larger segments and then the larger seg
ments are coupled to give the whole protein, the synthetic approac
involves the use of "cassettes." Thus, if one wishes to mak
structural analogs which involve, for example, the replacement of
particular secondary structural region in the protein, it is onl
necessary to synthesize a new variant of the particular segment(s
involved, but all the other segments employed to assemble the nativ
protein can be used for the mutant. In making a small protein b
the segment synthesis-condensation approach, the introduction o
non-peptidic segments like the one introduced in β-endorphin, i
very feasible as is specific substitution with labelled residue
useful for spectroscopic studies.

Once it is recognized that the segment synthesis-condensation ap
proach would be a very useful one for the construction of smal
proteins and their analogs, development of the chemistry allowing u
to perform these reactions with facility becomes very important.
significant advance in our ability to utilize the segment synthesis
condensation approach has been the development of the oxime este
polymer [52,54,55]. In brief, we start our syntheses with a p-ni
trobenzoyl oxime polymer constructed starting from polystyrene 1
crosslinked with divinylbenzene. The first amino acid with it
α-amino function and any reactive side chain groups protected i
attached using dicyclohexylcarbodiimide. Subsequently, the α-amin
function is deprotected, the next amino acid again in a suitabl
protected form is coupled, and the synthesis continues on the soli
support in much the same way as syntheses by the usual stepwis
solid-phase methodology (Figure 9). Removal of the growing peptid

53

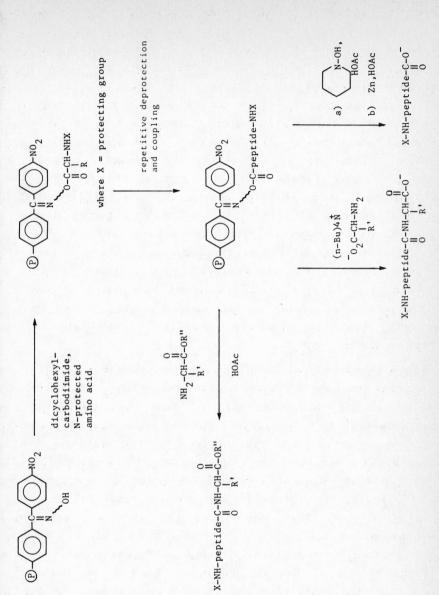

Figure 9. Synthesis of protected peptide segments by the oxime polymer method

chain from the typical benzyl ester bound polymeric form is achieved generally with vigorous conditions such as treatment with HF which results in deprotection of the side chain functional groups as well. In contrast, removal of the _protected_ peptide from the oxime polymer is achieved with facility under readily accessible conditions. For example, a growing peptide chain can be removed by the use of an amino acid ester and employing acetic acid as a catalyst to give a protected peptide segment which has the attacking amino acid group incorporated at the C-terminus. Also, the protected peptide segments can be removed in the free acid form by the use of a tetra n-butyl ammonium salt of an amino acid as the attacking group under quite mild acidic conditions [56]. Another approach for removal of the protected peptide segment from the resin which is very convenient involves the use of N-hydroxypiperidine. Reaction with this α-nucleophile gives rise to the protected peptide in the form of the O-acyl hydroxypiperidine species, which can be treated further with zinc and acetic acid, except if the segment contains a sulfur-containing amino acid, to give the protected peptide with a free C-terminal carboxyl function.

Using this methodology, we have constructed protected peptide segments covering the whole sequence of ribonuclease T_1 [57]. Our strategy involved dividing the molecule into three major portions which are illustrated in Figure 10. Each of these segments containing more than 30 amino acids was constructed from smaller protected segments. We have achieved the construction of the ribonuclease T_1 sequence in a fully protected form by coupling of the major segments. Similarly, a protected form of a structural analog of ribonuclease T_1 has been constructed in which the amphiphilic α-helical segment present in the N-terminal third of the molecule has been replaced by a designed α-helix which has greatly reduced sequence homology to the natural sequence. For both the natural and mutant enzymes, the protecting groups have been removed by the low HF/high HF treatment. Neither the natural enzyme nor the mutant have been purified to homogeneity. However, we have shown that in both cases the enzymes are catalytically active on yeast RNA [57]. Since the mutant enzyme has a completely redesigned amphiphilic α-helix this work shows that it is possible to replace in a systematic manner a whole secondary structural unit by a designed unit in an enzyme system, giving rise to a catalytic species with quite respectable activity. This represents a first step in the design of tertiary structure from primary amino acid sequence.

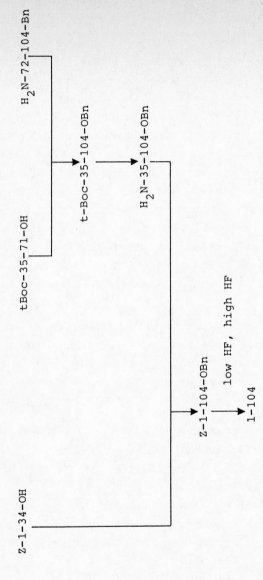

Figure 10. Synthetic strategy for the synthesis of ribonuclease T_1 by condensation of protected peptide segments

CONCLUSION

We have seen throughout this review that it is now quite feasible design peptide models for those biologically active molecules whic owe their biological and physical properties to the presence amphiphilic secondary structures in them. In a modest next step, have shown that we can extend the design principles elaborated fo the systems where tertiary structure could be neglected to example where tertiary structure is very important and where we have show it is possible for us to replace structural units and still maintai catalytic activity. In the future, we plan to pursue extensive the replacement of secondary structural features in a variety small proteins, including various enzymes, and the results of suc studies should certainly be useful in developing our ability t build tertiary structure from primary amino acid sequence.

ACKNOWLEDGEMENTS

The outstanding efforts of my coworkers whose names are cited in th various articles mentioned are gratefully acknowledged. Th research described here was supported in part by NIH Program Projec grant HL-18577.

REFERENCES
1 Kaiser ET (1970) Accounts Chem Res 3:145
2 Kaiser ET, Kaiser BL (1972) Accounts Chem Res 5:219
3 Baldwin RL, Eisenberg D (1987) In: Oxender DL, Fox CF Protei engineering. Liss, New York
4 Kaiser ET, Kezdy FJ (1983) Proc Natl Acad Sci USA 80: 1137
5 Kaiser ET, Kezdy, FJ (1984) Science 223:249
6 Ghosh SS, Bock SC, Rokita SE, Kaiser ET (1986) Science 231:145
7 Knowles JR (1987) Science 236:1252
8 Kaiser ET, Lawrence DS (1984) Science 226:505
9 Kaiser ET, Levine HL, Otsuki T, Fried HE, Dupeyre R-M (1980) Ad Chem 191:35
10 Polgar L, Bender ML (1966) J Am Chem Soc 88:3153
11 Neet KE, Koshland DE Jr (1966) Proc Natl Acad Sci USA 56:1606
12 Kuriyan J, personal communication
13 Segrest JP, Jackson RL, Morrisett JD, Gotto AM (1974) FEBS Let 38:247
14 Fitch WM (1977) Genetics 86:623
15 McLachlan AD (1977) Nature 267:465

5 Kroon DJ, Kupferberg JP, Kaiser ET, Kezdy FJ (1978) J Am Chem Soc 100:5975

7 Fukushima D, Kupferberg JP, Yokoyama S, Kroon DJ, Kaiser ET, Kezdy FJ (1979) J Am Chem Soc 101:3703

3 Fukushima D, Kaiser ET, Kezdy FJ, Kroon DJ, Kupferberg JP, Yokoyama S (1980) Ann NY Acad Sci 348:365

9 Yokoyama S, Fukushima D, Kupferberg JP, Kezdy FJ, Kaiser ET (1980) J Biol Chem 255:7333

0 Fukushima D, Yokoyama S, Kezdy FJ, Kaiser ET (1981) Proc Natl Acad Sci USA 78:2732

1 Fukushima D, Yokoyama S, Kroon DJ, Kezdy FJ, Kaiser ET (1980) J Biol Chem 255:10651

2 Nakagawa SH, Lau HSH, Kezdy FJ, Kaiser ET (1985) J Am Chem Soc 107:7087

3 DeGrado WF, Kezdy FJ, Kaiser ET (1981) J Am Chem Soc 103:679

4 DeGrado WF, Musso GF, Lieber M, Kaiser ET, Kezdy FJ (1982) Biophys J 37:329

5 Yates G, Tao HP, personal communication

6 Terwilliger TC, Eisenberg D (1982) J Biol Chem 257:6010

7 Terwilliger TC, Eisenberg D (1982) J Biol Chem 257:6116

3 Terwilliger TC, Weissmann L, Eisenberg D (1982) Biophys J 37:353

9 Colacicio G, Basu MK, Buckelew AR, Bernheimer AW (1977) Biochim Biophys Acta 465:378

0 Kapral FA (1972) Proc Soc Exp Biol Med 141:519

1 Whitelaw DD, Birkbeck TH (1978) FEMS Microbiol Lett 3:335

2 Taylor JW, Osterman DG, Miller RJ, Kaiser ET (1981) J Am Chem Soc 103:6965

3 Kaiser ET (1986) Ann NY Acad Sci 471:233

4 Taylor JW, Kaiser ET (1986) Pharmacol Rev 38:291

5 Moe GR, Miller RJ, Kaiser ET, (1983) J Am Chem Soc 105:4100

6 Moe GR, Kaiser ET (1985) Biochemistry 24:1971

7 Epand RM, Epand RF, Hui SW, He NB, Rosenblatt M (1985) Int J Pept Protein Res 25:594

8 Kaiser ET, Lynch B, Rajashekhar B (1985) Proc 9th Am Pept Symp, Toronto:855 Rockford IL, Pierce Chemical

9 Lau HSH, Rivier J, Vale W, Kaiser ET, Kezdy FJ (1983) Proc Natl Acad Sci USA 80:7070

0 Taylor JW, Miller RJ, Kaiser ET (1982) Mol Pharmacol 22:657

1 Taylor JW, Miller RJ, Kaiser ET (1983) J Biol Chem 258:4464

2 Blanc JP, Taylor JW, Miller RJ, Kaiser ET (1983) J Biol Chem 258:8277

43 Blanc JP, Kaiser ET (1984) J Biol Chem 259:9549

44 Rajashekhar B, Kaiser ET (1986) J. Biol Chem 261:13617

45 Musso GF, Assoian RK, Kaiser ET, Kezdy FJ, Tager HS (1984
    Biochem Biophys Res Commun 119:713

46 Musso GF, Kaiser ET, Kezdy FJ, Tager HS (1983) Proc 8th Am Pep
    Symp, Tucson:365, Rockford IL, Pierce Chemical

47 Velicelebi G, Patthi S, Kaiser ET (1986) Proc Natl Acad Sci US
    83:5397

48 Epand RM, Epand RF, Hui SW, He NB, Rosenblatt M (1985) Int
    Pept Protein Res 25:594

49 Musso GF, Kaiser ET, unpublished observations

50 Green FR Jr, Lynch B, Kaiser ET (1987) Proc Natl Acad Sci USA

51 Ikehara M, Ohtsuka E, Tokunaga T, Nishikawa S, Uesugi S, Tanak
    T, Aoyama Y, Kilyodani S, Fujimoto K, Yanase K, Fuchimura K
    Morioka H (1986) Proc Natl Acad Sci USA 83:4695

52 Clark-Lewis I, Aebersold R, Ziltener H, Schrader JW, Hood LE
    Kent SBH (1986) Science 231:134

53 Kaiser, ET Angewandte Chem Intl Ed, in press

54 DeGrado WF, Kaiser ET (1980) J Org Chem 45:1295

55 DeGrado WF, Kaiser ET (1982) J Org Chem 47:3258

56 Lansbury PT Jr, private communication

57 Sasaki T, Findeis M, personal communication

ALTERING THE STRUCTURE OF ENZYMES
BY SITE-DIRECTED MUTAGENESIS

W.H.J. Ward and A.R. Fersht

Department of Chemistry
Imperial College of Science and Technology
London SW7 2AY, UK

SUMMARY

Site-directed mutagenesis of enzymes allows allows the direct study of the contributions of specific side-chains to substrate binding and catalysis. This has led to a breakthrough in understanding relationships between structure and function. Tyrosyl-tRNA synthetase (TyrTS) from *Bacillus stearothermophilus* has been a paradigm for protein engineering studies on structure reactivity relationships in catalysis. Kinetic analysis of TyrTS mutants has given quantitative information which can be applied to many proteins. The interaction energy between the enzyme and its substrates during catalysis has been determined, allowing measurement of the apparent strengths of hydrogen bonds and salt bridges. It has been shown directly that catalysis results from stabilization of the transition state. The gross structure of TyrTS has also been investigated. The enzyme comprises 2 subunits of identical composition, and each monomer has 2 discrete functional domains: one which binds tRNA and the other which contains the active site. tRNA interacts with both subunits, binding to one and then being charged at the other subunit of the dimer. Each monomer has a complete active site, but only 1 site in each dimer functions catalytically. TyrTS is thus a classical example of an enzyme with half-of-the-sites activity. The mechanism of TyrTS has been studied and it has been shown that each dimer uses the same active site repeatedly. The second subunit has no detectable activity so that the enzyme has long-lasting asymmetry in function. Asymmetry is an inherent property and is not induced by binding of substrate. This accounts for half-of-the-sites activity and shows that the enzyme has an asymmetrical structure in solution, contrasting with the structure in crystals which is symmetrical about the subunit interface. A monomer of the enzyme is probably too small to allow both recognition and charging of tRNA, explaining the requirement for the enzyme to function as an asymmetric dimer. The enzyme appears to bind two molecules of Tyr sequentially to the same site during charging of 1 molecule of tRNA. The second molecule of Tyr perhaps aids the dissociation of charged tRNA by displacing the tyrosyl moiety from its binding site.

S. A. Benner (Ed.)
Redesigning the Molecules of Life
© Springer-Verlag Berlin Heidelberg 1988

GENETIC ENGINEERING HAS LEAD TO SUBSTANTIAL ADVANCES IN THE STUDY (PROTEINS

Recombinant DNA technology has revolutionized protein chemistry. Genes coding for speci proteins can be cloned, isolated, sequenced and expressed at high levels. The availability of a speci protein in large amounts removes many of the difficulties from purification. Site-directed mutagene: can be used to modify specifically a single side-chain thus allowing detailed study of relationshi between structure and function. Space does not permit a detailed review of the methodology a contribution of recombinant DNA technology to protein engineering, but this topic is cover elsewhere (1-4).

Site-Directed Mutagenesis is a Considerable Improvement upon Traditional Methods of Prote Modification.

Traditionally, the technique of chemical modification has been a cornerstone in studies relationships between protein structure and function. Now, site-directed mutagenesis has render chemical modification studies largely obsolete. Given the resources to perform site-direct mutagenesis, then, in general, it improves over the old technique in many ways:

1) *Specificity*. A single residue may be modified.

2) *Lack of side-reactions*. This is a common problem in chemical techniques.

3) *Global applicability*. Any residue can be modified genetically whereas chemical modificati usually requires accessibility to solvent.

4) *Flexibility*. Any naturally-occurring amino-acid may be introduced whereas chemical approach generally increase the volume of the side-chain by adding a substituent group. Observed changes function may, therefore, be due to loss of reactivity or simply due to steric effects. However, unli chemical modification, protein engineering does require availability of a cloned, isolated and sequenc gene together with a suitable expression system. Further, it is essential to purify the enzyme of inter away from any enzymes that have the same activity and which are endogenous to the expressi system.

Protein engineering allows relationships between structure and function to be dissected in f detail. However, such application of site-directed mutagenesis requires structural information about enzyme under investigation. Chemical modification studies can be a very efficient approach in order give structural information prior to protein engineering (see ref 5). Structural data can also be obtain by a number of other techniques including from X-ray crystallography, NMR spectroscopy homologies in amino-acid sequence.

INTRODUCTION TO THE AMINOACYL-tRNA SYNTHETASES

Immediately before the development of site-directed mutagenesis, the catalytic mechanism was r

own for any of the aminoacyl-tRNA synthetases. Extensive investigations using protein chemistry
d enzyme kinetics had failed to identify groups involved in catalysis. Crystal structures of the
zymes gave no immediate clues as to catalytic mechanism. In the last 5 years, protein engineering of
e TyrTS has given more information on structure-function relationships, catalytic mechanism and
ergetics than we know about most enzymatic reactions. Quantitative dissection of enzyme structure
d activity has provided much general information on enzyme catalysis and molecular recognition in
dition to answering specific questions about this important class of enzymes.

zyme Catalysis as a Series of Free-Energy Changes.

The basis of enzyme catalysis is the employment of binding energy between enzyme and
bstrate in order to increase rate and specificity when compared to the uncatalysed reaction free in
lution. Thus, to understand enzyme catalysis we must know the magnitude of interaction energies
tween the enzyme and its substrate throughout the whole course of the reaction. This allows study
how binding energy is used to lower activation energy, change unfavourable equilibrium constants
d determine specificity. The following basic strategy has been used to analyse relationships
tween structure and function in TyrTS:
Potentially important residues were identified by X-ray crystallography.
Selected side-chains were changed by oligonucleotide-directed mutagenesis.
The mutant proteins were characterised quantitatively using enzyme kinetics.

rTS is an Excellent System for Protein Engineering

TyrTS from *B. stearothermophilus* has proved to be exceptionally suitable for protein
gineering studies. High resolution (0. 25 nm) X-ray structures for the enzyme are available (see
low). The gene has been cloned, sequenced and expressed at high levels, in *E. coli* (6,7). Further,
ncentration of active enzyme may be determined accurately by active-site titration (8) enabling
lculation of thermodynamic parameters from kinetic measurements. Stable intermediates on the
ction pathway accumulate on a suitable time-scale for analysis by pre-steady state kinetics (9). This
ilitates detailed study of enzyme mechanism. The enzyme is very stable under experimental
nditions since it originates from a thermophilic bacterium. Furthermore, this stability allows TyrTS
ivity from the *E. coli* expression system to be abolished simply by heating preparations of *B.
arothermophilus* enzyme. This procedure does not affect the enzyme from the thermophile and so
asured kinetic properties relate directly to the enzyme of interest.

TyrTS crystallizes as a symmetrical dimer, with each subunit having a complete active site. Each
nomer has two domains: the subunits interact through the N-terminal domains and the C-terminal
mains are not involved in contacts between the monomers (10). Aminoacylation of tRNA is
alysed as a two-step reaction (11).

$$E + Tyr + ATP \rightleftarrows \quad E\ Tyr\text{-}AMP + PPi \qquad\qquad (1)$$

$$E.Tyr\text{-}AMP + tRNA \rightarrow \quad E + Tyr\text{-}tRNA + AMP \qquad\qquad (2)$$

First, Tyr is activated by formation of E.Tyr-AMP, which is stable in the absence of tRNA. The stability is critical for the functioning of the enzyme since it changes the equilibrium constant in favour of formation of Tyr-AMP and also sequesters the reactive intermediate which would otherwise diffuse from the enzyme and either hydrolyse or modify nucleophiles (12). The kinetics of activation of Tyr are easily measured by monitoring isotope exchange from radiolabelled PPi into ATP at chemical equilibrium (13). This is known as the PPi-exchange assay. In the second step (eq 2), activated Tyr is transferred to tRNA. Deletion of the C-terminal domains of the enzyme generates ΔTyrTS which cannot bind tRNA, but has little effect on the kinetics of the activation reaction (14; Figure 1).

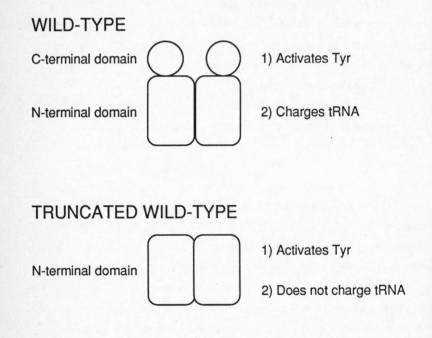

Figure 1. Functional Domains in TyrTS

The great stability of the E.Tyr-AMP complex allows determination of concentration of active enzyme by active site titration (8). Enzyme is mixed with radioactive Tyr and ATP in the presence of pyrophosphatase. E.Tyr-AMP accumulates fully since the equilibrium (eq 1) favours products due drolysis of PPi. The radiolabelled complex is collected on nitrocellulose filters which are then shed to remove unbound label, and then assayed by scintillation counting. X-ray crystallography s given high resolution structures of the free enzyme, E.Tyr complex and E.Tyr-AMP (10; Figure . Further, the structures of ΔTyrTS, both as a free dimer, and complexed with Tyr have been cidated (15). The C-terminal domains of the full-length enzyme are not sufficiently ordered to give a fined electron density in crystals (10). However, the structures of the N-terminal domains in the ld-type and truncated enzymes are very similar, correlating with the kinetic properties of the two zymes (14, 15). Wild-type TyrTS crystallizes as a symmetrical dimer with the axis of symmetry ssing along the subunit interface. Each subunit has a complete active site (10). Crystals of truncated zyme show some asymmetry, but, like the full length enzyme, are largely symmetrical about the bunit interface (15).

gure 2. Interactions Between Tyr-AMP and Side-Chains of TyrTS. Hydrogen bonds suggested by alysis of the crystal structure of E.Tyr-AMP are shown by dashed lines.

EASUREMENT OF INTERACTION ENERGY BETWEEN ENZYME AND SUBSTRATE

ermodynamic Characterization of TyrTS

Detailed kinetic analysis of TyrTS mutants has allowed measurement of the strength of eractions between the catalyst and its substrate throughout the course of the activation of Tyr. easured rate and binding constants have been used to calculate the free-energy change at each step of e reaction

Pre-steady-State Kinetics has Given Essential Information on the Mechanism of TyrTS

In traditional steady-state kinetics, the enzyme turns over many times during the course of the assay. The rate of change of concentration of substrate or product is measured, thus the rate and binding constants determined are only *apparent* values since individual steps may not be dissected from the overall mechanism. Intermediates cannot be detected directly using steady state kinetics and so this approach gives only very limited information about enzyme mechanism. However measurement of *absolute* rate and binding constants (that is those for individual steps on the reaction pathway) during the first turnover of the enzyme (in the pre-steady state) allows detection of intermediates as they are formed and then decay during the course of the enzyme reaction. Pre-steady state kinetic analysis is, therefore, required in order to obtain essential information on reaction mechanism. TyrTS is ideally suited for study in the pre-steady state since the intermediate, E.Tyr-AMP, is very stable and its rate of formation is sufficiently slow to be measured. Further, most processes in the pre-steady state follow first-order rate constants. This enables very accurate measurements to be made since the observed half-time is independent of [E] added to the assay. However, the validity of pre-steady state data can only be confirmed after comparison with steady-state measurements. Here again, TyrTS is a good system to study since both the activation of Tyr and charging of tRNA can be assayed in the steady state. Activation of Tyr is followed using the PPi exchange assay, the radiolabelled PPi that is formed is collected by adsorption to activated charcoal followed by filtration. Aminoacylation of tRNA is monitored using $[^{14}C]$Tyr, precipitating charged tRNA using acid and then filtering in order to collect the precipitate.

Pre-steady state assays usually require sophisticated and specialized equipment in addition to very large amounts of enzyme. This is because events on the enzyme are being observed directly so that there is no amplification due to multiple turnovers as in a steady state experiment. However, cloning of the TyrTS gene and its expression in *E. coli* have provided sufficient enzyme for pre-steady state work.

Kinetic Characterization of TyrTS in the Pre-Steady State.

Each of the rate or binding constants in the activation of Tyr (Figure 3) can be measured accurately either using pre-steady state kinetics or equilibrium dialysis (9). Pre-steady state kinetics are generally followed by monitoring fluorescence change using a stopped-flow spectrometer. Mutant enzymes where the rate constant for the catalytic step is drastically reduced may be assayed in the pre-steady state simply by monitoring the time course of active site titration. From steady-state kinetics activation of Tyr appears follows a random order of addition of substrates (Figure 3). However, the crystal structure of E.Tyr-AMP (10) suggests that binding of ATP would prevent access to the Tyr binding site and so the reaction is probably ordered with E $\rightarrow$ E.Tyr $\rightarrow$ E.Tyr.ATP. This does not

fect steady-state kinetics as the system behaves as if equilibrium is rapidly reached between E.ATP
d E.Tyr.ATP via the free enzyme.

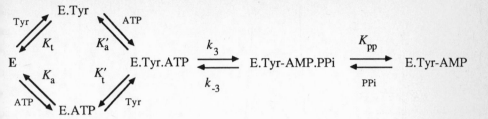

gure 3. Kinetic scheme for the activation of tyrosine

The kinetic characteristics of TyrTS in the pre-steady state are determined as follows. K_t, the
ssociation constant of Tyr from E.Tyr can be measured by equilibrium dialysis or kinetics. K'_a, the
ssociation constant of ATP from E.Tyr.ATP, and k_3 the rate constant for formation of
Tyr.AMP.PPi may both be measured by stopped-flow on mixing E.Tyr with ATP. k_{-3}, the rate
nstant for pyrophosphorolysis and K_{pp}, the dissociation constant of PPi from the E.Tyr-AMP.PPi
mplex may also be measured by stopped-flow on mixing E.Tyr-AMP with PPi. (The E.Tyr-AMP
mplex may be prepared by mixing E with Tyr and ATP, incubating and then purifying from
reacted substrate by gel-filtration. The complex is very stable, hydrolysing with a half-life of several
urs.) K_a, the dissociation constant of ATP from E.ATP, and K'_t, the apparent dissociation
nstant of Tyr from E.Tyr.ATP can be measured from steady state kinetics. The rate constant for
arging of tRNA can also be measured using stopped-flow fluorescence by mixing E.Tyr-AMP with
NA (11).

alculation of the Magnitude of Free-Energy Changes from Measured Rate and Binding Constants

The free-energy changes during an enzyme catalysed reaction may be summarized on a reaction
ordinate diagram (Figure 4). First, binding energy, ΔG_s, is released on formation of the E.S
mplex. In order to be converted into product, the substrate must pass through the transition state
hich is the most unstable complex on the reaction pathway. The change in free-energy required to
ach the transition state is the activation energy ($\Delta G^\ddagger$) and represents the thermodynamic barrier
hich substrate must cross before conversion into product. Passage over this barrier is, therefore, the
te determining step. The free-energy changes during the course of the reaction may be calculated
om measured rate and binding constants:

$$G_s = RT \ln K_s \qquad (3)$$
$$G^\ddagger = -RT \ln k_{rds} + RT (\ln k_B T/h) \qquad (4)$$

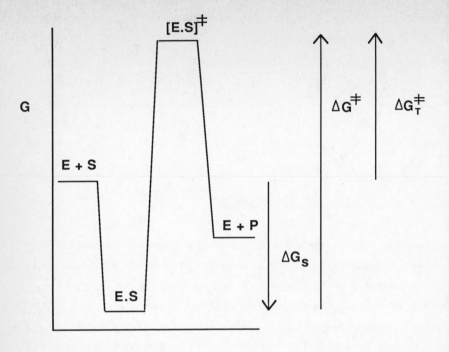

Figure 4. Reaction coordinate diagram for the enzyme catalysed reaction.

Where K_S is the dissociation constant of the E.S complex, k_{rds} is the rate constant of the ra[te] determining step, k_B is the Boltzmann constant and h is the Planck constant. The apparent activati[on] energy, or the change in free-energy in going from free E to transition state, is given by summing eq[ns] and 4, thus:

$$\Delta G_T^{\ddagger} = -RT \ln(k_{rds}/K_S) + RT \ (\ln k_B T/h) \qquad (5)$$

Note that the change in binding energy with the transition state resulting from a mutation is given by:

$$\Delta\Delta G_T^{\ddagger} = -RT \ \ln[(k_{rds}/K_S)_{mut}]/[(k_{rds}/K_S)_{wt}] \qquad (6)$$

where the subscripts *mut* and *wt* denote mutant and wild-type enzymes respectively. The appare[nt] binding energy of a given side-chain may thus be measured by deleting the group by site-direct[ed] mutagenesis and then comparing the kinetic characteristics of the mutant enzyme with those of t[he] wild-type. Detailed pre-steady state analysis is usually required to measure absolute values of K_S a[nd] k_{rds}. Thermodynamic characterization of TyrTS is greatly simplified since, for ATP dependen[ce] pyrophosphate exchange,

$$k_{cat}/K_s = k_{cat}/K_m \qquad (7)$$

his makes measurement of interaction energies relatively simple for mutants of TyrTS.

Each equation from 3 to 6 gives the free-energy change under standard thermodynamic conditions 1 M with respect to all reagents. All free-energy changes quoted in this work refer to these onditions since this allows direct comparison with any other result determined under standard onditions.

STRATEGY FOR PROTEIN ENGINEERING

Design of Mutant Enzymes

Mutation of the enzyme may have three effects.

Loss of interactions from the original side-chain.

Gain of modified interactions from the new side-chain.

Structural reorganization, which may be delocalised throughout the enzyme.

is wise, therefore, to describe values determined in this manner of experiment always as *apparent binding energies*. Interpretation of the physical significance of apparent binding energies is greatly mplified if the mutation is carefully designed in order to minimise effects 2) and 3) above. This ondition is likely to be satisfied if the following guidelines are utilized.

Change the amino-acid residue to one which is *smaller* than that in the wild-type.

Delete a single functional group *without* introducing a new one.

Analysis of Mutant Enzymes.

Having made a suitable mutation, it is then essential to check that delocalized structural organization has not occurred otherwise the physical significance of measured free-energy changes ill remain unknown. The following approaches have been used when studying TyrTS.

Kinetic analysis. Measure all rate and binding constants for the mutant enzymes. Any changes om the wild type should be predictable from the X-ray structure. Clearly pre-steady state values are ore reliable than those from the steady state since changes in individual steps may offset each other ading to no detectable effect in the steady state.

Structural analysis. This approach has been accelerated by using difference maps to compare ectron density maps determined by X-ray crystallography of mutant and wild-type TyrTS (16).

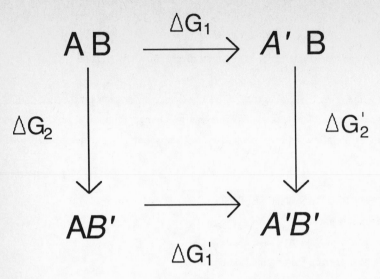

Figure 5. Energetic changes in the double-mutant test (17). A and B are mutated to A' and B'.

3) *Double-mutant test*. Free-energy changes for double mutants should be additive when compared t[o] the parental single mutants (17). Consider mutating 2 residues, A and B, in the wild-type to produc[e] the double mutant, $A'B'$ (Figure 5). If no structural rearrangement occurs, then the effect of mutatin[g] A to A' is the same whether the second residue is B (ΔG_1) or B' ($\Delta G'_1$). Likewise, the effect [of] changing B to B' is independent of whether the second residue is A (ΔG_2) or A' ($\Delta G'_2$). Thus, [if] replacement of each side-chain induces no structural reorganization, their effects will be independe[nt] and the overall change in interaction energy in double mutant will be the sum of the correspondi[ng] terms for the two single mutants, and so eq 8 will hold.

$$\Delta G_1 = \Delta G'_1, \Delta G_2 = \Delta G'_2 \qquad\qquad (8)$$

If one of the mutations causes structural reorganization, then the effects will not be additive and eq [8] will not hold. A change of a few hundredths of a nm in a strong hydrogen bond would correspond t[o] 8-13 kJ/mol and so would be readily detectable as a 25-200 fold change in k_{cat}/K_m. However, suc[h] a change would be very difficult to detect by X-ray crystallography. But, apparent additivity ma[y] result from fortuitous cancelling out of the effects of structural mutations and so care is required whe[n] interpreting these experiments. Further, this approach only detects interactions between specific pai[rs] of residues, it does not give an overview of the whole structure of the mutant enzyme. Large structur[al] changes may thus remain undetected simply because a pair of residues do not interact.

X-ray crystallography suggests that residues Cys-35 and His-48 of TyrTS form hydrogen bon[d]

ith 3'-hydroxyl and the ring oxygen in the ribose of Tyr-AMP (10; Figure 2). Application of the
ouble mutant test implies that changing these residues to glycines does not lead to gross reorganization
 the enzyme structure since $\Delta G_1 = \Delta G'_1$ and $\Delta G_2 = \Delta G'_2$ (Figure 6).

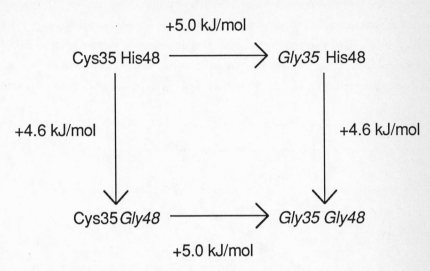

igure 6. Application of the double-mutant test to changes at positions 35 and 48.

Linear free-energy relationships. Consider the equilibrium between E.S and E.P, a systematic
ariation in the reactivity of should produce a linear relationship between activation energy and
quilibrium free-energy. This is an application of the Brønsted or Hammett relationship of physical
rganic chemistry. This approach has been employed in protein engineering as a sensitive test for
elocalized structural reorganization on changing the structure of the enzyme rather than the substrate
8, 19). The Brønsted theory predicts that a family of mutants should fit to a single linear free-energy
elationship (LFER) when comparing activation energy and equilibrium free-energy. The theoretical
asis of LFERs is as follows. Assume that the relationship between equilibrium constant, K, and rate
onstant, k, is given by:

$$= AK^\beta \tag{9}$$

here A and ß, are both constants. The plot of ln$k$ against ln$K$ when eq 9 holds is a straight line of
ope ß, (eq 10).

$$k = \beta \ln K + \text{constant} \tag{10}.$$

k is related to the activation energy ($\Delta G^\ddagger$ eq 4) and ln$K$ is proportional to the equilibrium free energy (ΔG_{eq}).

$$\Delta G^{\ddagger} = \text{ß}\Delta G_{eq} + \text{constant} \tag{11}.$$

A value of ß, close to 0 means that the transition state resembles starting materials, and a value close
1 means that the transition state resembles products. This behaviour applies when the residu
concerned show a uniformly progressive change in binding energy as the reaction proceeds. Wher
whole series of mutants can be fitted to a single linear free-energy equation, there is very stro
evidence that each mutant shows part of a general trend in the relationship between structure a
activity.

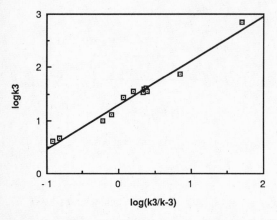

Figure 7.   Linear free energy plot for k_3 and $K = k_3/k_{-3}$ (in Figure 3) for the formation of
E.Tyr-AMP from E.Tyr.ATP.

Wild-type enzyme together with 11 mutants at positions mainly affecting binding of the ribo
ring have been found to fit to a Brønsted plot for the step E.Tyr.ATP ⇌ E.Tyr-AMP.PPi (Figure
indicating that none of these changes causes delocalized structural reorganization of the enzyme (1
The ß-value obtained is 0.8 indicating that the transition state for this step is quite similar to t
products. Further, each step in the activation of Tyr can be analyzed in this way simply by consideri
the rate and equilibrium constants of the appropriate part of the pathway. Thus, the 12 enzymes
several Brønsted plots showing that, on average and relative to 100% for binding Tyr-AMP, 12%
the binding energy of these interactions is realized on binding of ATP to the E.Tyr complex, 84%
reaching the transition state [Tyr-ATP]‡ and 91% in the E.Tyr-AMP.PPi complex (19). This sho
how the binding energy of apparently non-catalytic residues contributes to altering k_3 and t
equilibrium constant between E.Tyr plus ATP on one side of the equation and E.Tyr-AMP plus PPi
the other side.

HYDROGEN BONDING AND BIOLOGICAL SPECIFICITY

When compared with other types of bonds, hydrogen bonds are uniquely important in relationships between protein structure and function. This is because hydrogen bonds are sufficiently strong to direct specificity, but sufficiently weak to be made and broken during the course of catalysis. Before the development of protein engineering, there was little evidence on the strengths of these interactions in solution. There are complicating effects caused by water being the solvent and water being a hydrogen bond donor and acceptor itself. Binding energies represent the *differences between the ligand and receptor bound to water, and the ligand and receptor bound to each other*. The importance of complementary hydrogen bonds in catalysis and molecular recognition has been measured quantitatively using protein engineering of TyrTS (20).

The crystal structure of the E.Tyr-AMP complex (Figure 2) suggested a large number and variety of hydrogen bonds exist between the enzyme and the intermediate. The initial strategy was to mutate hydrogen bonding side-chains to remove donors or acceptors and so produce mutants lacking specific bonds. Comparison of the kinetics of activation of Tyr (measured using the pyrophosphate exchange assay) by wild-type and mutant enzymes (eq 6 and 7) would then give empirical determinations of the contribution of hydrogen bonding to specificity and catalysis (Table 1).

Table 1: Apparent Binding Energies for Individual Side-Chains of TyrTS.

Enzyme group	Mutant	Substrate group	$\Delta\Delta G^{\ddagger}_T$ (kJ mol^{-1})
Tyr34-OH*	Phe	Tyr-OH	2.2
Cys35-SH*	Gly	ATP 3'-OH	5.0
Cys51-SH	Ala	ATP ring O	2.0
His48-imidazole NH*	Gly	ATP ring O	4.6
Asn48-amide NH$_2$	Gly	ATP ring O	3.2
Cys35-SH*	Ser	ATP ring O	4.9
Tyr169-OH*	Phe	TyrNH$_3^+$	16.0
Gln195-NH$_2$	Gly	TyrCOO$^-$	19.0

*Indicates wild-type TyrTS. Side-chains on the enzyme were changed by protein engineering in order to leave unpaired hydrogen bond donors or acceptors on the substrate. The change in apparent activation energy, $\Delta\Delta G^{\ddagger}_T$, measures the apparent binding energy, ΔG_{app}, of the deleted group under standard thermodynamic conditions of 1 M with respect to each reagent. Data from ref. 20.

Deletion of a side-chain between enzyme and substrate to leave an unpaired, uncharged hydrogen bond donor or acceptor weakens binding by 2-5 kJ/mol. This change in apparent binding ener, (ΔG_{app}) increases by some 12 kJ/mol when there is an unpaired and charged donor or accepte However, the physical significance of ΔG_{app} is often difficult to interpret, especially when mutati leaves a charged donor or acceptor unpaired (21, 22). Consider the situation where a hydrogen bo acceptor, -B, on a substrate forms a bond with a hydrogen bond donor, -XH, on the enzyme.

$$S\text{-}B\cdots HOH + H_2O\cdots HX\text{-}E = [S\text{-}B\cdots HX\text{-}E] + H_2O\cdots HOH \qquad (12)$$

Insight into the net energetics can be gained by simply counting the number and types (that is, wheth or not one of the partners is charged) of hydrogen bonds on both sides of eq 12, that is performing *hydrogen bond inventory*. This gives only an estimate of the free-energy change expected since does not allow for the possibility that hydrogen bonds of the same type may differ in strength. Th reaction is approximately isoenthalpic because the are the same number and types of bonds on bo sides of eq 12. However, hydrogen bond formation in enzyme-substrate interactions is energetical' favourable because there is a gain in entropy on release of enzyme-bound and substrate-bound wat into bulk solution.

Deleting a hydrogen bond to an uncharged donor or acceptor weakens binding by 2-5 kJ/m which is considerably less than the calculated strength of hydrogen bonds *in vacuo* (13-27 kJ/m (23)). The hydrogen bond inventory (see eq 13) shows that there is no net change in number or typ of hydrogen bonds during the reaction, as was also the case when a paired hydrogen bond w: formed.

$$S\text{-}B\cdots HOH + H_2O\ \ E = [S\text{-}B\ \ E] + H_2O\cdots HOH \qquad (13)$$

Poorer binding on deletion of -XH is largely due to small changes in solvation energies and possib losses of dispersion energy in the complex because a hole is present between S and E, which th missing group on the side-chain once occupied.

Consider a charged group on the substrate which can form a hydrogen bond with a side-chain i the wild-type enzyme

$$S\text{-}B^{-}\cdots HOH + H_2O\cdots HX\text{-}E = [S\text{-}B^{-}\cdots HX\text{-}E] + H_2O\cdots HOH \qquad (14)$$

The hydrogen bond inventory again shows no net change in number or type of hydrogen bonds c formation of E.S complex. Deletion of a bond to a charged donor or acceptor weakens binding b approximately 15-20 kJ/mol. On formation of the E.S complex, there is again no change in number o hydrogen bonds, but a charged bond is exchanged for an uncharged one:

$$3^- \cdots HOH + H_2O \quad E = [S-B^- \ E] + H_2O \cdots HOH \qquad (15)$$

is accounts for part of the change in ΔG_{app}, but much of the effect is probably caused by S-B⁻ ving to form compensating interactions when XH is deleted. These interactions are likely to cause ictural reorganization in the protein.

Protein engineering of TyrTS has, therefore, allowed investigation of the strengths of hydrogen nds. Similar results have been found for the removal of bonding groups from the substrate in order perturb charged and uncharged hydrogen bonds in complexes between glycogen phosphorylase and ar (24). Quantitative measurements have shown that an individual uncharged hydrogen bond is not reat determinant of biological specificity since the presence of a single unpaired, uncharged, drogen bond donor or acceptor in a complex increases its dissociation constant only by a factor of 2 20. The presence of a number of unpaired groups will amplify this specificity ratio. The presence an unpaired but charged donor or acceptor, however, can weaken binding by several orders of gnitude. The most important factors in specificity are steric repulsion and unsolvated charges at the erfaces of complexes.

:ASUREMENT OF THE STRENGTHS OF ENGINEERED SALT-BRIDGES

ermodynamic parameters can readily be calculated from kinetic analysis of TyrTS. The enzyme can s be used to study relationships between structure and function which apply to any protein. The bility of charged groups in proteins was, therefore, investigated by construction of TyrTS mutants. vo charged side-chains were engineered into symmetrical positions at the subunit interface of both l-length, and truncated, enzyme (Figure 8). When the residues have the same polarity, the dimers sociate reversibly into monomers which cannot bind, nor activate, Tyr (25-27). Dissociation of d-type TyrTS has never been detected. Mixing mutants so that the residues at position 164 have posite polarity generates heterodimers, and not homodimers, indicating that a single salt-bridge is ficient to direct specificity in subunit association (26,27).

The target for mutagenesis was Phe-164, a residue which is on the symmetry axis of crystalline yme such that the side-chains of the two Phe-164 residues interact in a very hydrophobic region. -164 was changed to Asp or Glu in full-length enzyme and to Lys or Arg in truncated enzyme (27). size-difference allowed subunit association to be studied by gel-filtration chromatography. sociation from active homodimers into inactive monomers is favoured when pH induces ionization position 164. Mixing of mutants of opposite charge generates heterodimers and monomers, but no ectable homodimers. Equations were derived in order to analyze the kinetics of any enzyme which

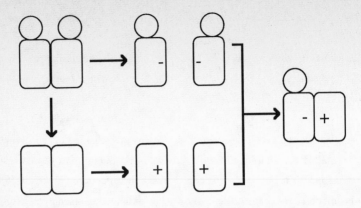

Figure 8. Effects of introducing charged residues into the subunit interface of TyrTS and ΔTyrT
Negatively charged side-chains were introduced into the full-length enzyme, and positively charg
residues were introduced into the truncated enzyme. Note that dissociation is favoured when sid
chains at position 164 have the same polarity, whereas association is favoured when these side-chai
have opposite polarity.

dissociates reversibly into inactive subunits (27). The basis of the analysis is as follows. Substr
binds only to dimeric enzyme, therefore, dissociation of the enzyme decreases affinity for the liga
causing an increase in the value of K_m that is measured. (The dimeric forms of mutants at positi
164 of TyrTS have values of K_m similar to those of the wild-type, whereas monomers cannot bi
substrate and so have an infinite value of K_s.) The value of K_m is determined and then used
calculate the proportion of dimer that has dissociated and thus the dissociation constant (K_d) of
dimeric enzyme.

Application of this analysis to mutants at position 164 allowed measurement of values of K_d
the dimers. The values obtained are absolute equilibrium constants and are thus independent
concentration of enzyme or substrate. The free-energy change on association of subunits und
standard thermodynamic conditions can, therefore, be calculated directly from the dissociati
constant. The heterodimers have K_d values of 6-14 μM, whereas for mutant homodimers K_d = 1:
4000 μM. Replacement of Phe-164 by fully ionized residues in mutant homodimers strongly favo
dissociation, mainly because of the repulsion between charges having the same polarity. Mut
dimers are much less stable than the wild-type even when the two residues at position 164 ha
opposite polarity which could attract the subunits to each other. Salt-bridges might be expected to
very stable in a hydrophobic environment because of the strong electrostatic interactions. Howev
the salt bridges engineered into the heterodimers are relatively weak since the heterodimers read
dissociate and dissociation of wild-type enzyme has never been detected. The salt bridges are
practice, destabilized because of the high self energy of the charged groups - they are effectiv
desolvated. Naturally occurring salt-bridges are stronger than this because their charges are stabili

✓ complementary interactions with neighbouring polar groups in the protein. The artificial salt-
ridges were engineered into TyrTS are not stabilized in this way. Nevertheless, a single salt-bridge is
efficient to direct specificity in subunit association.

MEASUREMENT OF FREE-ENERGY CHANGES DURING CATALYSIS

Analysis using steady-state kinetics has identified residues which are involved in substrate binding and
measured their contributions. The next step was to measure how the strengths of these interactions
changes as the reaction proceeds. These changes in free-energy can be interpreted in terms of the
mechanism of catalysis. Free-energy profiles for the activation of Tyr by wild-type and mutant
enzymes have been constructed using pre-steady state kinetics and equilibrium binding to measure rate
and binding constants. The apparent contribution of the binding energy of a side-chain which had been
mutated, ΔG_{app}, was determined by subtracting the energy level of each complex of the wild-type
enzyme from that of the mutant enzyme and then visualized by plotting as a *difference energy diagram*
(28, 29). Changes in ΔG_{app} as the reaction proceeds from one intermediate to the next on the reaction
pathway, $\Delta \Delta G_{app}$, can provide a direct measure of the changes in interaction energy of a particular
side-chain with the substrate. Interaction energies throughout the activation reaction are summarized in
Table 2, and the deduced stereochemical interactions in the transition state are shown in Figure 9. The
side-chains binding Tyr and ATP may be classified functionally as follows.

Tyr-Binding Site: Specificity in Amino-Acid Binding

Residues Asp-78, Tyr-169 and Gln-173 form a binding site for the α-amino group of Tyr.
Mutation of any one of these side-chains weakens binding by about 13 kJ/mol. The interaction energy
of Tyr-169 remains unchanged throughout the reaction, but the other residues are more difficult to
analyze as they are involved in hydrogen bond networks which are disturbed upon mutation (30). The
site which determines specificity for Tyr rather than Phe is composed of Asp-176 and Tyr-34. The
most important group is the carboxylate of Asp-176 which functions as a hydrogen bond acceptor for
the substrate hydroxyl. Unfortunately, mutation of this residue has never produced active enzyme.

TyrTS Catalyses Activation of Tyr by Preferential Stabilization of the Transition State

Haldane (31) and Pauling (32) proposed that enzymes catalyse reactions by stabilizing the
transition state. The basis of the theory is as follows. Enzymes bind their substrates by
complementary interactions. However, the structure of the substrate changes during the course of the
reaction. It begins at the ground state and must pass through the transition state before conversion into
products. It is catalytically advantageous for the enzyme to have the highest degree of complementarity
with the most unstable complex on the reaction pathway (that is the transition state) since this lowers
$G^{\ddagger}$ and so increases k_{rds}. The enzyme would, therefore, be expected to show *differential binding*,

Table 2. Interaction energies of side chains throughout reaction.

Residue	Interaction energy of side chains in complex with:				
	Tyr	ATP	[Tyr-ATP]‡	PP$_i$	Tyr-AMP
Tyrosine binding site					
Tyr-34	+	0	+	0	+
Asp-78	++++	++*	++++	++*	++++
Tyr-169	++++	0	++++	0	++++
Gln-173	++++	++*	++++	+*	++++
Nucleotide and pyrophosphate site					
Cys-35	0	0	++	0	+++
Thr-40	0	0	++++	++++	0
His-45	0	0	++++	++++	0
His-48	0	0	+++	0	+++
Thr-51	0	0	0	0	-
Lys-82	0	++	++++	++++	0
Arg-86	0	0	++++	++++	-
Asp-194	0	0	++++	+	+++
Lys-230	0	0	++++	++++	0
Lys-233	0	++++	++++	++++	0

Key: Values of $\Delta\Delta G_{app}$ on mutation: 0 = -2 to +2; +, = 2 to 4; ++ = 4 to 6; +++ = 6 to 8; ++++ = >8; - = -2 to - 4 kJ/mol. * = evidence for some disruption of protein structure on mutation.

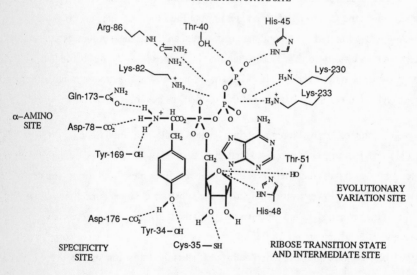

Figure 9. Interactions of the tyrosyl-tRNA synthetase with the the transition state for the formation of Tyr-AMP

vouring the transition state over the ground state. The increase in binding energy on moving into the nsition state can then be used to increase rate of reaction above that in free solution.

Before the development of protein engineering, compounds that resemble the proposed transition ate (transition state analogues) were found to bind more tightly to enzymes than do native substrates e ref 33). However, the relationship between structure and binding of the authentic transition state uld not be directly investigated. Protein engineering of TyrTS has provided direct evidence for nsition state stabilization leading to increased catalytic rate.

At the start of this study, the structure of the transition state was inferred from the structure of Tyr-AMP. Kinetic analysis of mutant enzymes was then used to construct difference energy agrams which provided direct evidence of transition state structure. Not only have difference energy agrams quantified the importance of interactions suggested by the crystal structure, but they have also ad to the discovery of unexpectedly important residues.

bstrate Motion Allows Stronger Binding in the Transition State

Protein engineering has demonstrated that a principal catalytic factor in the activation of Tyr by rTS is improved binding of ATP when moving from the E.Tyr.ATP complex into the transition state 4). The transition state was modelled by adding the PPi group to the crystal structure of E.Tyr-MP. This approach suggested that the γ-phosphate moves into a binding pocket on the enzyme when e tyrosyl carboxylate attacks the tetravalent α-phosphate of ATP to generate a pentacoordinate insition state. The side-chains of Thr-40 and His-45 appeared to bind the PPi moiety in the transition ate. Removal of these groups by protein engineering shows that they do not bind free ATP, but teract strongly with the transition state. Accordingly, the mutant TyrTS(Thr→Ala-40; His→Gly-45) s the value of k_3 lowered by 3.2×10^5 but the value of K'_a is raised by only a factor of 5. The fference energy diagrams for the mutants at positions 40 and 45 provide clear evidence for their nding ATP in the transition state and PPi in the E.Tyr-AMP.PPi complex. Mutation has little effect the E.Tyr, E.Tyr.ATP and E.Tyr-AMP complexes but considerably weakens binding in the E.[Tyr-TP]‡ and E.Tyr-AMP.PPi complexes.

zyme Motion Allows Stronger Binding in the Transition State.

Random mutagenesis (35) revealed a set of residues which were crucial in catalysis, but whose les were not immediately obvious as they are either remote from the reagents or so disordered that ey were initially difficult to localize. These residues are positively charged groups which have now en shown to be in mobile loops of the enzyme that move in order to envelope the negatively charged i moiety during the transition state in an induced-fit mechanism (36). Residues Lys-82 and Arg-86, ich are on one rim of the binding site pocket, and Lys-230 and Lys-233, which are on the other side, have been mutated to Ala residues. The mutants have values of k_3 which are reduced up to 80 fold. Construction of difference energy diagrams reveals that all of the residues interact specifica with the transition state and with PPi in the E.Tyr-AMP.PPi complex. Modelling of the transition sta using the structure of E.Tyr-AMP places the ε-NH$_2$ groups of Lys-230 and Lys-233 at least 0.8 n too far away to interact with the PPi moiety in the transition state at the same time as do Lys-82 a Arg-86 (Figure 9). Binding of substrates must, therefore, induce a conformational change in t enzyme that brings these residues into range. Consistent with this proposal is the observation that four residues are in regions of the protein that have high crystallographic temperature factors, whi indicate flexibility.

Catalytic Mechanism for Activation of Tyr.

Protein engineering has shown directly that TyrTS catalyses formation of Tyr-AMP stabilization of the transition state. Movements in both the substrate and the enzyme allo strengthening of interactions when moving from the ground state to the transition state. Th mechanism appears to involve use of binding energy with no classical general acid-base or covale catalysis. Catalysis seems to be delocalized over the whole binding site. The reaction is summarized Figure 10. In the ground state (top left), E.Tyr.ATP complex, ATP binds to Lys-82 and Lys-23 There is insignificant binding energy with Cys-35, Thr-40, His-45, His-48, Thr-51, Lys-82 Arg- and Lys-230. In the transition state (top right), these groups all interact either with the PPi moiety with the ribose ring of ATP. In the E.Tyr-AMP.PPi complex (bottom left), the groups still intera with the reagents. In the E.Tyr-AMP complex (bottom right), the Lys-230/Lys-233 loop moves aw from the adenylate after dissociation of the PPi, and Lys-82 binds to the α-phosphate of Tyr-AMP a bridging water molecule.

TyrTS HAS TWO IDENTICAL SUBUNITS, BUT IS ASYMMETRICAL IN SOLUTION

TyrTS crystallizes as a symmetrical dimer, with each subunit having a complete active site (10). Ea of these sites can function since the crystalline enzyme can bind and activate 2 mol Tyr/mol dimer (1 37). However, in solution, the enzyme binds tightly, and activates rapidly only 1 mol Tyr/mol dim (8, 38). The enzyme is, therefore, said to display half-of-the-sites activity. This has been conside to be an example of extreme negative cooperativity since binding of substrate to one subunit w thought to inhibit binding at the second subunit of the dimer. However, recent studies (39) show th half-of-the-sites activity is not a cooperative phenomenon, but instead is caused by inherent asymmet in the dimer.

Half-of-the-sites activity occurs in a number of enzymes (see ref 33), including most aminoac tRNA synthetases (40). However, in most cases, both the mechanism and the function of th phenomenon are unclear. Protein engineering has provided a powerful approach in studying half-c

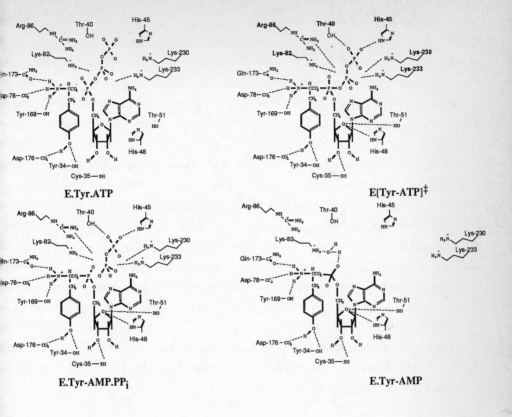

Figure 10. Steps in the formation of E.Tyr-AMP from E.Tyr.ATP.

the-sites activity in TyrTS. Heterodimers, comprising two subunits of different composition have be constructed allowing introduction of different mutations into each subunit of the dimer. Heterodime produced by mutagenesis at the subunit interface are difficult to use in study of half-of-the-sites activi because they are in equilibrium with inactive monomers. However, fully stable heterodimers ha been produced by reversible denaturation of a mixture of full-length and truncated enzymes in urea (3 41).

The full-length subunit of heterodimers binds tRNA which is charged using Tyr activated at t truncated subunit (41). This was shown using the mutation His→Asn-45 which decreases the value k_3 by a factor of 10^4. If this mutation is on the full-length subunit, then the heterodimer can char tRNA. If the change is on the truncated subunit, the enzyme can activate Tyr but is unable to char tRNA. Detailed kinetic analysis of heterodimers suggests that wild-type TyrTS has long-lastir asymmetry in solution (39). This is because dimers repeatedly use the same active site. The secor site has no detectable activity. The enzyme is, thus, frozen in two populations: 50% are active at or subunit and 50% at the other. For example, heterodimers containing the Asn-45 mutation on or subunit form 0.5 mol of Tyr-AMP/mol dimer rapidly at the wild-type rate and a further 0.5 mol ve slowly at the rate expected for mutant (Figure 11). Any interconversion of active and inactive subun is thus on a much slower time-scale than the several minutes half-life for formation of Tyr-AMP at t mutated site.

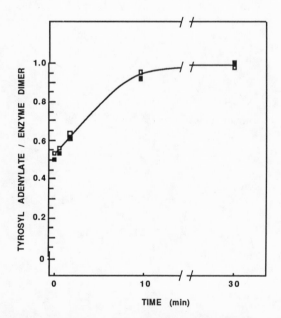

Figure 11. Time-Course for formation of Tyr-AMP by heterodimers containingthe mutatic His→Asn-45 at one active site. The mutation was either in the full-length subunit (open symbols) the truncated subunit (filled symbols).

Asymmetry is maintained when the enzyme turns over in the steady-state. A heterodimer of full-
ngth wild-type enzyme combined with truncated wild-type enzyme charges tRNA at half of the rate of
e full-length wild-type dimer. This is because half of the enzyme in solution forms Tyr-AMP at the
ll-length subunit and so remains as an inactive dead-end complex which is unable to charge tRNA,
hereas the other half of the enzyme turns over at the truncated subunit, charging tRNA at the wild-
pe rate (Figure 12). If this half of the enzyme in solution were to use the active site on the full-length
bunit, then it would also become inactive dead-end complex. This complex is very stable and so
ould accumulate leading to a decrease in the rate of charging as a function of time. No such decrease
 seen indicating that the active site on the truncated subunit is used repeatedly. Enzyme-substrate
mplexes for wild-type TyrTS are asymmetrical in solution because half-of-the-sites activity is
hibited. The results of the work on heterodimers suggest that the free enzyme is also asymmetrical
nce there is no evidence that behaviour of heterodimers is different from that of wild-type enzyme
9). Further, pre-steady-state kinetics indicate that wild-type TyrTS has identical properties for the
rst turnover and all subsequent turnovers (9), confirming that no cooperative effect occurs and so free
zyme must be asymmetrical in solution. Crystalline TyrTS is symmetrical about the subunit interface
d so is different in structure from enzyme in solution. However, the crystalline enzyme is able to
tivate Tyr (36) and so must be very similar in structure to the enzyme in solution. Protein
gineering has, therefore, detected a significant difference in structure between the enzyme in solution
d that in crystals. Crystallization possibly selects a symmetrical conformation of TyrTS which is too
re to be detected in solution. TyrTS does not display negative cooperativity, but instead has inherent,
ng-lasting asymmetry in structure which leads to half-of-the-sites activity.

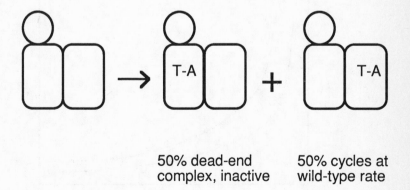

50% dead-end
complex, inactive

50% cycles at
wild-type rate

igure 12. Charging of tRNA by a heterodimer of full-length wild type combined with truncated wild
pe. In the first turnover, half of the population forms Tyr-AMP (T-A) on the full-length subunit and
e other half of the population uses the truncated subunit. Enzyme which uses the truncated subunit
r the first turnover then uses that subunit repeatedly for subsequent turnovers when charging tRNA.

MECHANISM OF tRNA CHARGING

Binding of tRNA.

A heterodimer of full-length and truncated enzyme provides an asymmetric system f•
identification of groups involved in binding tRNA. This enabled Bedouelle & Winter (35) to assig
by random mutagenesis, patches of Lys and Arg residues which bind tRNA and this has enabled a lo•
resolution model of the E.tRNA complex to be built.

Catalysis of Charging.

TyrTS shows half-of-the-sites activity in formation of Tyr-AMP which results from long-lastir
asymmetry of the enzyme. Paradoxically, however, the kinetics of tRNA charging are biphasic wi•
respect to [Tyr] (11) suggesting that both subunits of the dimer can be catalytically active in th
presence of tRNA. This paradox has now been resolved by investigating the mechanism of tRN
charging (42).

Biphasic kinetics could result from catalysis at the second active site only in the presence of tRN.
and high [Tyr]. Utilization of both active sites could thus lead to charging of 2 mol tRNA during eac
turnover. This proposal was disproved by deleting 1 of the 2 tRNA binding domains. The resulta•
heterodimer retains biphasic Tyr-dependence of charging (Figure 13) showing that the second phase •
not caused by binding a second mol of tRNA.

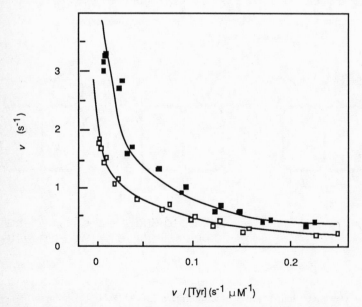

Figure 13. Tyr-dependence of charging of tRNA. Eadie plot for wild-type enzyme (filled symbol•
and a heterodimer of full-length wild type combined with truncated wild type (open symbols). No•
that each enzyme gives a curved plot because it diplays 2 different values of k_{cat} and K_m.

Binding of two mol Tyr during aminoacylation of a single mol tRNA could also produce biphasic kinetics. One subunit could catalyse charging whilst the other subunit of the dimer binds Tyr, releasing energy which is used to increase catalytic rate. However, kinetic analysis of enzymes containing different mutations in each subunit of the dimer shows only binding of Tyr to the catalytic subunit affects the observed rate of reaction (42). TyrTS, therefore, maintains half-of-the-sites activity when charging tRNA. Biphasic kinetics appear to result from two molecules of Tyr binding sequentially to the same site. The first molecule would have to dissociate from the binding site before the end of the reaction pathway is reached leaving the site empty to allow access of a second molecule of Tyr. This model is consistent with the long-lasting asymmetry of the enzyme since only one subunit binds substrate. Alternatively, biphasic kinetics could simply result from division of the enzyme into two populations which have different kinetic properties. This possibility is very unlikely, especially since neither activation of Tyr nor ATP-dependence of charging are biphasic

The reaction may proceed as follows. The Tyr moiety of Tyr-tRNA could leave its binding site and the product could remain associated with the enzyme via the tRNA. This would leave the Tyr-binding site accessible to a second mol of the substrate. This model is supported by the observation that TyrTS can bind two moles of radiolabel in the presence of [14C]Tyr-tRNA and [14C]Tyr (11).

A number of aminoacyl-tRNA synthetases display half-of-the-sites activity and biphasic kinetics with respect to the cognate amino-acid (40). For some of these enzymes, it has been shown that binding of a second mol of amino-acid increases flux through the rate limiting step in aminoacylation which is dissociation of charged tRNA (40). TyrTS may utilize additional substrate binding in this way. It seems likely that binding a second mol of amino-acid displaces charged tRNA from the amino-acid binding site and so increases the rate of dissociation of the product. This may be a common mechanism amongst aminoacyl-tRNA synthetases. These studies also suggest why TyrTS has half-of-the-sites activity. tRNA is a large molecule which binds to one subunit and is then charged using Tyr activated at the other subunit of the dimer (35, 40, 42). It appears necessary for the enzyme to function as an asymmetric dimer in order to be large enough to both recognize and charge tRNA.

EVOLUTION OF ENZYME STRUCTURE AND ACTIVITY

The fine tuning of enzyme catalysis by evolution can be studied by directed mutagenesis. Although the mutants so made are not necessarily the evolutionary precursors of the present wild-type enzyme, they do exemplify the types of changes that can occur. We have shown the changes postulated by Albery & Knowles (43) do occur (29). Further, theories on the evolution of values of K_M by Fersht (44) are found to hold (45). The very existence of linear free energy relationships is important in theoretical work on evolution.

ELECTROSTATIC EFFECTS IN CATALYSIS

Electrostatic effects are extremely important in structure and catalysis. These are ubiquitous in th
folding, assembly and interaction of macromolecules, and ligand binding. They have crucial effects i
catalysis by stabilizing charged transition states and modulate the pH dependence of catalysis. It is n
an easy matter to calculate or quantitate electrostatic effects in proteins because of the uncertainty
dielectric constant - the protein is itself heterogeneous and there are mathematical problems arisin
from the water/protein boundary protein with a sharp transition from high to low dielectric media. Th
problem of measuring and manipulating long-range electrostatic effects in proteins was tackled usin
subtilisin as a model system. The activity of this archetypical serine protease from *Bacillu*
amyloliquefaciens follows the ionization of a histidine at the active site (His-64). The pK_a of th
residue may be accurately measured by kinetics. The gene was cloned and expressed in *Bacillu*
subtilis (46, 47). The strategy was to mutate charged surface residues. This perturbed the pK_a
His-64 and it was possible to measure with accuracy the shift in pK_a. From this, the electrostat
interaction energies between surface charges and His-64 were readily calculated. In all case
irrespective of whether there is protein or water between the two charges, the effective dielectr
constant is found to be high and approaches that of water (48). The data so derived was used to te
methods of calculating electrostatic interactions. The Warwicker-Watson algorithm was successful i
reproducing the data to some 10-20% accuracy (49).

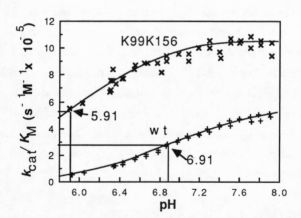

Figure 14. Electrostatic effects on the pH dependence and catalysis of subtilisin. Two negatively
charged residues which are 1.3-1.5 nm from His-64 (Asp-99 and Glu-156) were mutated t
positively-charged lysine residues. This lowers the pK_a of His-64 by 1 unit and causes the enzyme t
be more active at lower pH. The substrate is negatively charged and so the increase in surface charg
of the enzyme increases the enzyme-substrate affinity.

As well as providing the basic information on long-range electrostatic interactions, this work i
important in protein design because it affords a general procedure for manipulating the pH dependenc

f enzyme activity, specificity towards charged substrates and even catalysis itself (Figure 14) (46, 8). This also opens up a new field of enzyme chemistry - studying the charge distribution in ionic ansition states - which is reminiscent of the use of substituent effects in physical-organic chemistry.

JTURE PROSPECTS

Protein engineering of TyrTS has elucidated important principles in relationships between tucture and function. Much of this information can be applied widely to different proteins and so can : used to produce desired changes in properties of proteins. This information can be used in design specific inhibitors and, ultimately, in the design of new enzymes *de novo* in order to catalyse any emical reaction.

EFERENCES

Leatherbarrow RJ, Fersht AR (1986) Protein Engineering 1:7-16
Carter P (1986) Biochem J 237:1-7
Knowles JR (1987) Science 236:1252-1258
Shaw WV (1987) Biochem J 246:1-17
Ward WHJ (1987) Trends Biochem Sci 12:28-31
Winter G, Fersht AR, Wilkinson AJ, Zoller M, Smith M (1982) Nature 299:756-758
Winter G, Koch GLE, Hartley BS, Barker DG (1983) Eur J Biochem 132:383-387
Fersht AR, Ashford JS, Bruton CJ, Jakes R, Koch GLE, Hartley BS (1975) Biochemistry 14:1-4
Fersht AR, Mulvey RS, Koch GLE (1975) Biochemistry 14:13-18
Blow DM, Brick P (1985) In: Jurnak F, McPherson A (eds) Biological Macromolecules and Assemblies Nucleic Acids and Interative Proteins, Wiley, New York 2:442-469
Jakes R, Fersht AR (1975) Biochemistry 14:3350-3356
Wells TNC, Ho C, Fersht AR (1986) Biochemistry 25:6603-6608
Calendar R, Berg P (1966) Biochemistry 5:1681-1690
Waye MMY, Winter G, Wilkinson AJ, Fersht AR (1983) EMBO J 2:1827-1830
Brick P, Blow D (1987) J Molec Biol 194:287-297
Brown KA, Brick P, Blow DM (1987) Nature 326:416-418
Carter P, Winter G, Wilkinson AJ, Fersht AR (1984) Cell 38:835-840
Fersht AR, Leatherbarrow RJ, Wells TNC (1986) Nature 322:284-286
Fersht AR, Leatherbarrow RJ, Wells TNC (1987) Biochemistry 26:6030-6038
Fersht AR, Shi JP, Knill-Jones J, Lowe DM, Wilkinson AJ, Blow DM, Brick P, Carter P, Waye MMY, Winter G (1985) Nature 314:235-238
Fersht AR (1987) Trends Biochem Sci 12:301-304
Fersht AR (1987) Biochemistry 26:8031-8037
Weiner SJ, Kollman PA, Case DA, Singh UC, Ghio C, Alagona G, Profeta S, Weiner P (1984) J Amer Chem Soc 108:765-784
Street IP, Armstrong CR, Withers SG (1986) Biochemistry 25:6021-6027
Jones DH, McMillan AJ, Fersht AR, Winter G (1985) Biochemistry 24:5852-5857
Ward WHJ, Jones DH, Fersht AR (1986) J Biol Chem 261:9576-9578
Ward WHJ, Jones DH, Fersht AR, (1987) Biochemistry 26:4131-4138

28   Wells TNC, Fersht AR (1986) Biochemistry 25:1881-1886

29   Ho C, Fersht AR (1986) Biochemistry 25:1891-1897

30   Lowe DM, Winter G, Fersht AR (1987) Biochemistry 26:6038-6043

31   Haldane JBS (1930) Enzymes, Longmans Green & Co Ltd, UK (Reprinted 1965 MIT Press Cambridge, USA)

32   Pauling L (1946) Chem Eng News 24:1375-1377

33   Fersht AR (1985) Enzyme structure and Mechanism, Freeman, New York

34   Leatherbarrow RJ, Fersht AR, Winter G (1985) Proc Natl Acad Sci, USA 82:7840-7844

35   Bedouelle H, Winter G (1985) Nature 320:371-373

36   Fersht AR, Knill-Jones J, Bedouelle H, Winter G (1987b) Biochemistry (in press)

37   Monteilhet C, Blow DM (1978) J Molec Biol 122:407-417

38   Fersht AR (1975) Biochemistry 14:5-12

39   Ward WHJ, Fersht AR (1988) Biochemistry (in press)

40   Schimmel P, Soll D (1979) Ann Rev Biochem 48:601-648

41   Carter P, Bedouelle H, Winter G (1985) Proc Natl Acad Sci, USA 83:1189-1192

42   Ward WHJ, Fersht AR (1988) (submitted for publication)

43   Albery WJ, Knowles J R (1976) Angew Chem Int Ed Engl 16: 285-293

44   Fersht AR (1974) Proc R Soc London B187:397-407

45   Fersht AR, Wilkinson AJ, Carter P, Winter G (1985) Biochemistry 24:5858-5861

46   Thomas PG, Russell AJ, Fersht AR (1985) Nature 318:375-376

47   Russell AJ, Thomas PG, Fersht, AR (1987) J Mol Biol 193:803-813

48   Russell AJ, Fersht AR (1987) Nature 328:496-500

49   Sternberg MJE, Hayes FR, Russell AJ, Thomas PG, Fersht AR (1987) Nature 330:86-88

Structure and Activity of Ribozymes

J.W.Szostak

pt. of Molecular Biology, Massachusetts General Hospital, and Dept.
of Genetics, Harvard Medical School, Boston, MA 02115 USA

STRACT

In this article I will review the catalytic activities of the
cently discovered RNA enzymes, or ribozymes. I will emphasize the
oup I introns, since much more is known about this group of self-
licing introns than any other ribozyme. These molecules are of
rticular interest because the transesterification reaction they
talyze is chemically the same as the reaction that a primitive RNA
plicase would have to carry out. The little information that we have
out the structure of these RNA enzymes comes from phylogenetic
quence comparison data, although detailed genetic analysis promises
be quite informative. I will also discuss the use of nucleotide
alogues and base modification procedures as structural probes. After
iefly reviewing the other known and possible ribozymes, I will
nclude with some speculative comments on the possibilities of making
w enzymes, both from RNA and from novel ordered heteropolymers.

TRODUCTION

The discovery just 5 years ago by Cech and his coworkers that RNA
lecules could have catalytic activity (1,2) has had a strong impact
thinking about the origin and evolution of catalysis and of life
self. Prior to this discovery, it was generally assumed that only
oteins had the structural and chemical diversity necessary for
nerating the broad range of specific catalysts required to make up a
ving organism. This assumption, however, led to a fundamental
radox: the production of proteins is a complex process which, in all
own organisms, requires the participation of many proteins. It
emed impossible that such a process could evolve spontaneously.
deed, it had been proposed that RNA must have played an essential

S. A. Benner (Ed.)
Redesigning the Molecules of Life
© Springer-Verlag Berlin Heidelberg 1988

catalytic role in early living systems, since only nucleic acids cou
potentially be both information carriers and active, structurally
complex catalysts (3-6). However, these proposals were not widely
taken seriously in the absence of any examples of RNA enzymes.

Now, on the basis of a handful of ribozymes that catalyze
essentially two reactions (hydrolysis and transesterification of
phosphodiester bonds), speculation has run wild in the opposite
direction. Proposals abound of a complex 'RNA world', in which cells
had evolved a complex metabolism mediated entirely by RNA enzymes (7
9). The many nucleotide cofactors have been proposed to be a relic o
that era. These ideas lead directly to an interesting but difficult
question: what range of chemical reactions could be efficiently
catalyzed by RNA enzymes? Few ribozymes seem to have survived the
evolutionary onslaught of protein catalysts, and it may therefore se
that this question must remain forever unanswered. However, techniqu
for the synthesis of RNA molecules are advancing so rapidly that it
may soon be possible to design new ribozymes in the laboratory.

The first reports that the splicing of the Tetrahymena ribosomal
RNA intron proceeded without the help of any protein were greeted wi
considerable skepticism. It was generally assumed that some tightly
bound catalytic protein must have copurified with the RNA, perhaps i
very small amounts. Definitive proof that the RNA was self-splicing
was quickly obtained, however, by the in vitro synthesis of the RNA
transcription of a cloned DNA template by E. coli RNA polymerase (2)
These purified transcripts, which had never seen a Tetrahymena (or a
other) cell, were as active as the molecules isolated from
Tetrahymena.

The ability to rapidly synthesize large amounts of RNA by in vit
transcription has allowed extensive biochemical and genetic
characterization of this and other catalytic RNAs. Such RNA synthese
are now generally accomplished with the aid of the bacteriophage SP6
or T7 RNA polymerases, which are small stable single subunit enzymes
with very high activity (10,11). Run-off transcripts are made on a
restriction enzyme digested DNA template. The transcripts begin insi
the RNA polymerase promoter sequence, and end at the end of the DNA
template. Manipulation of the plasmid template by standard recombina

techniques allows the rapid generation of enzyme sequence
iants.

Since the initial proof by Cech's laboratory that the Tetrahymena
zyme was a catalytic RNA, three other classes of RNA enzymes have
en discovered. The tRNA processing enzyme RNase P is a
onucleoprotein, and it was known that the RNA component was
ential for activity (12). Shortly after the intron work, it was
nd that it was the RNA component of this enzyme that was in fact
 catalytically active subunit (13). This enzyme differs from the
f-splicing intron in two fundamental ways: it catalyzes a
drolysis reaction, not a transesterification, and it acts in trans
 a separate substrate molecule, a tRNA precursor. More recently, it
 been found that certain plant pathogens known as viroids and
usoids are small RNAs that self-cleave as a normal part of their
uration pathway (14,15). This reaction is a hydrolysis reaction
t occurs in cis. Finally, a separate class of introns, unrelated to
 Tetrahymena intron, also has some self-splicing members (16,17).
se are more complicated enzymes that catalyze a series of
nsesterification reactions, one of which generates a 2'-5' linkage.
 is notable that all four well characterized catalytic RNAs act on
 substrates. It is possible that RNA is particularly well adapted
 the catalysis of reactions on an RNA substrate because of the
plicity of sequence recognition (by Watson-Crick base pairing). If
 was ever able to catalyze reactions involving other classes of
strates, those enzymes have apparently been replaced by protein
ymes, presumably because of their greater efficiency.

 GROUP I INTRONS

The group I introns are a set of introns defined on the basis of a
mon secondary structure and certain conserved features of their
leotide sequences (18). Although they comprise only a very small
ction of known introns (most introns in nuclear genes are spliced
 a snRNP mediated pathway), they are extremely widely distributed
logenetically, being found in some bacteriophages (19), chloroplast
) and mitochondrial (21) genes, and even a few nuclear genes (22).
logenetic comparison data, to be discussed in more detail below,
 been a very useful source of structural information about these
ecules.

Not all of the group I introns are self-splicing. For example, several introns from Neurospora mitochondria appear to interact with mitochodrial tyrosine tRNA synthetase, which acts as an essential cofactor (23). In many other group I introns, self-splicing has not been observed, and the additional factors required for activity rema unknown. In other cases, splicing is either not observed or occurs a very low rates in vitro under roughly physiological conditions, but can be readily detected at higher temperatures and increased Mg^{++} concentration. Indeed, the Tetrahymena intron, by far the best studi of all the group I introns, self-splices about 50 fold faster in viv than in vitro, suggesting the existence of specific factors that interact with the RNA enzyme and increase the rate of splicing (2). vivo, unspliced precursor ribosomal RNA cannot normally be detected, suggesting a half life on the order of seconds. In vitro, under optimized conditions, the half time of the reaction is about 1 to 2 minutes. Although slow compared many protein enzymes, this rate of turnover is roughly comparable to that of many restriction enzymes.

Catalytic Activities

Very little is known of the structure of this enzyme, or the mechanism of catalysis. The intact intron, embedded in its natural flanking exon sequences, catalyzes its excision from the ribosomal P with remarkable precision. These precise cleavage ligation reactions and the absolute requirement for guanosine as a cofactor, show that highly specific nucleotide recognition must play a key role in the functioning of this enzyme. Perhaps even more remarkable than the highly specific recognition is the range of reactions that the enzym can catalyze when its normal flanking sequences are removed. Althoug these reactions are almost certainly not physiologically significant they are of great chemical and evolutionary interest.

Transesterications: The normal pathway of autocatalytic splicing involves a set of three sequential transesterification reactions. Ea of these reactions involves the breaking of a specific phosphodieste bond and the concerted formation of a new phosphodiester bond. The first of the three reactions is the cleavage of the bond between the last base of the 5' exon and the first base of the intron, with concommitant formation of a bond between a free guanosine and the

st base of the intron. The 3′ hydroxyl of the free guanosine
acks the phosphate linking the exon and the intron and the
nosine displaces the intron, presumably by a standard in line
leophilic substitution reaction. In the second reaction, the 3′
droxyl of the displaced exon attacks the phosphate linking the
ron and the 3′ exon, with the result that the 5′ exon displaces the
ron and becomes joined to the 3′ exon, generating the intact
duct ribosomal RNA. In the final step, the 3′ hydroxyl of the
placed intron attacks an internal site in the intron, displacing a
rt oligonucleotide and forming a circular version of the intron.
chemical similarity of these three steps is matched by the
ilarity of potential secondary structures involving the interacting
leotides (24,25). Thus, it seems reasonable to view the whole
icing process as a cascade of three highly similar reactions
alyzed by a common catalytic domain in the RNA, presumably
responding to the phylogenetically conserved core of the intron.

ns Splicing: In order to test this hypothesis directly, I carried
a deletion analysis of the sequences required for splicing
ivity (25). The intron and flanking sequences were cloned
nstream of a T7 RNA polymerase promoter. A series of truncated
plates was generated by digestion of the DNA with restriction
ymes that cut either in the 3′ exon, or at various distances inside
3′ end of the intron. Transcription of these templates generated
molecules differing at their 3′ ends: all transcripts that
tended beyond the central, conserved portion of the intron retained
ability to self-cleave at the 5′ exon-intron junction.
roximately 80 nucleotides at the 3′ end of the intron are therefore
necesary for activity. A second series of modified templates was
n constructed by making internal deletions across the 5′ exon-
tron junction. Although lacking any natural substrate, and therefore
active in cis, these transcripts retained the ability to act in
ns, and could cleave added substrate RNAs.

cific hydrolysis: Under certain conditions, the Tetrahymena intron
catalyze the hydrolysis of specific phosphodiester bonds (26).
is reaction was first observed during analysis of the circular RNA
med during the final splicing reaction. This circular RNA is slowly
verted to a linear form, by hydrolysis at the phosphodiester bond
the site of circularization. That particular phosphate is therefore

specifically activated, perhaps by being in a strained conformation
due to interactions with the enzyme. Hydrolysis generates a 3'
hydroxyl and 5' phosphate. This is similar to the transesterificatio
reactions, which always leave a 3' hydroxyl, but quite different fro
standard RNA cleavage reactions (by RNases or alkali) which result i
a 5' hydroxyl and a 2'-3' cyclic phosphate. The hydrolysis reaction
pH dependent, and is first order with respect to hydroxide ion. The
simplest interpretation of this reaction is that hydroxide ion can
attack the activated phosphate, in the absence of the 3' hydroxyl of
the nucleotide that would normally do so.

The linearized intron can undergo a second round of autocatalyti
circularization, again by a transesterification reaction, this time
accompanied by release of a tetranucleotide fragment. This smaller
circle is again subject to linearization by hydrolysis. The final
linear form, referred to as L-19 RNA because it is 19 nucleotides
shorter than the original excised intron, does not recircularize, bu
does retain catalytic activity for certain additional reactions.

It has subsequently been found that both the 5' and 3' splice
sites of the intact precursor RNA are also susceptible to hydrolysi
(27). Hydrolysis of the 5' splice site appears to occur at a lower
rate than for the 3' site. However, this may be because the 5' exon
generated by hydrolysis is a substrate for the second splicing
reaction, exon ligation, and is therefore rapidly consumed. Since al
phosphodiester bonds that are substrates for transesterification are
activated for hydrolysis, a common mechanism is probably involved. O
reasonable possibility is that the enzyme distorts the phosphate gro
away from its normal tetrahedral geometry towards the trigonal
bipyramidal geometry that is the likely transition state for a
nucleophilic substitution reaction, thus lowering the activation
energy for both reactions.

Sequence specificity: What determines which phosphates are activate
for transesterification or hydrolysis? The answer is fairly clear in
the case of the 5' exon-intron junction, and rather less so in the
case of the intron-3' exon junction and circularization sites. The
last nucleotide of the upstream exon is always a U for group I
introns, and this base always occurs as a U:G base pair embedded in
base-paired stem. One side of the stem consists of sequences flankin

e exon-intron junction, and the other side is a sequence within the
tron called the internal guide sequence (24). The highly conserved
ture of the stem and the U:G base pair suggest that the enzyme
ecifically recognizes this structure and activates the phosphate 3'
 the U residue. The base paired stem has been shown to be essential
r catalysis by genetic experiments: single mutants on either strand
at disrupt the pairing decrease splicing activity, and double
tants that restore pairing restore activity (28,29). The essential
ture of the U:G base pair has been established in the trans-splicing
stem described above (Doudna and Szostak, in preparation). Mutant
bstrates were prepared in which the U:G base pair was changed to all
her possible base pairs - only U:G was active. Furthermore, the
sition of the U:G could be moved up and down within the stem and
licing always occurred 3' to the U, suggesting that the U:G was both
cesary and sufficient for determining the site to be attacked by the
 hydroxyl of the guanosine.

   Selection of the 3' splice site appears to be more complex. The
st nucleotide of the intron is always a G, and this G appears to
cupy the same binding site as the free G involved in the first
licing reaction. Some evidence in favor of this hypothesis is
scussed below. Davies et al. (24) originally proposed that the 3'
on would pair with the internal guide sequence, in such a way as to
come aligned correctly relative to the 5' exon. However, this
iring is possible only in about half of the group I introns.
reover, deletion of the internal guide sequence or the normal 3'
on exposes cryptic acceptors within the intron that do not fit this
mple rule (30,25). Site selection for circularization is similarly
mplex, and additional experiments will be required to clarify this
pect of the specificity of the intron.

 RNA restriction enzyme: The 'exon-intron junction' and the
ternal guide sequence must be complementary, but need not be on the
me molecule. Thus, the L-19 RNA, which retains the internal guide
quence, can bind other RNA molecules that contain short regions
mplementary to the IGS, and cleave them by guanosine attack. Thus,
e L-19 RNA can act as an RNA restriction enzyme (31), with a Km for
s RNA substrate of about 1 uM. Cleavage always occurs 3' to a U in
e substrate RNA that lies at the end of a 4 nucleotide stretch
mplementary to the IGS. Since cleavage is mediated by guanosine

attack, the 3' fragment of the cleaved substrate begins with the add
G. The specificity of cleavage for a tetranucleotide exactly
complementary to the IGS can be enhanced by performing the reactions
in 2.5 M urea or 15% formamide, which presumably destabilizes the
binding of mismatched sequences. Mutants of the IGS prevent binding
of sequences that are substrates for the wild type enzyme, but allow
different complementary substrates to bind and be cleaved. Thus an
appropriate panel of mutant RNAs should be able to cleave any
tetranucleotide RNA sequence ending in U. This reaction is also of
interest because it is clearly catalytic: multiple turnover is readily
observed, with a kcat of 0.13/min.

Nucleotidyl transferase and RNA polymerase: Perhaps the most
remarkable reactions yet observed with the Tetrahymena enzyme involve
short oligonucleotide substrates (32). When supplied with
pentacytidylic acid (C5), the enzyme generates both shorter and longer
oligo-C products. The reaction mechanism begins with binding of the
oligo-C to the poly-purine region of the internal guide sequence of
the L-19 enzyme, where it is subject to attack by the 3' hydroxyl of
the terminal G of the intron. The resulting transesterification
reaction leads to transfer of one or two C residues from the C5
oligomer to the 3' end of the intron, and release of the remainder of
the oligo-C chain. The modified intron is a covalent enzyme:substrate
intermediate. In the second step of the reaction, a new oligo-C chain
binds to the internal guide sequence, and the reverse reaction occurs
ie. transfer of the terminal C residues of the intron to the oligo-C
chain. The net result is the disproportionation of two oligo-C
molecules into one longer and one shorter molecule. Repeated cycles
reaction can generate long (>30)chains of oligo-C. Thus, the enzyme
acting as an RNA polymerase, in which the template is the internal
guide sequence, and the substrate is C5.

If the template was external instead of internal, it is easy to
imagine the enzyme acting as an RNA replicase. Cech (33) has proposed
a model for a primordial RNA replicase based upon the activity of the
L-19 enzyme.

Phosphotransferase and phosphatase activity: The L-19 enzyme
described above can also act as a highly specific phosphotransferase
and acid phosphatase (34). When oligo-C substrates ending in a 3'

osphate were prepared, the enzyme was able to transfer the 3'
osphate from the oligonucleotide to the terminal G of the intron,
d transfer it back to another oligo-C substrate with a 3' hydroxyl.
ly oligonucleotides with sequences complementary to the internal
ide sequence were substrates for this reaction, suggesting that one
tive site is common to all reactions. Although it is surprising that
e enzyme can act on a phosphomonoester (all of its other reactions
e on phosphodiesters), Zaug and Cech suggested that at the low pH
timum of 5 for this reaction, the 3' phosphate would be protonated
d could react similarly to a diester. Specific hydrolysis of the 3'
osphate on the enzyme is also observed at low pH.

ructure of Group I introns

ylogenetic Analysis: What little information we have concerning the
 structure of the self-splicing group I introns comes largely from
ylogenetic sequence comparison analysis and genetic analysis,
pplemented to some extent by data from chemical probes. The great
ylogenetic diversity of the group I introns has already been
ntioned. Given the large apparent evolutionary distances separating
e existing group I introns, the extent of conservation of the
imary and secondary structures of these enzymes is quite striking.
ring and Davies first pointed out the existence of four short
retches of sequence homology common to all group I introns (24).
ese consensus sequences allow different intron sequences to be
igned. Potential secondary structures can then be generated, either
nually or by computer. Although a very large number of structures
pear possible for any one sequence, only one structure is consistent
th all of the available sequences. A given base paired region is
garded as proven if, in several closely related sequences,
mpensatory base changes are found. Based on this type of evidence,
ry similar models for the secondary structure of the group I introns
re proposed independently by three groups (24,35,36). Since that
me, many additional group I introns have been sequenced (over 40 by
w), and all contain the original consensus sequences and fit the
andard secondary structure. Recently, most researchers in the field
ve agreed upon a standard format and nomenclature for the group I
tron structure (37). The base-paired stems are labeled P1 through
, as they occur in order from 5' to 3'.

Comparison of the secondary structures of a large number of group I introns reveals astonishing conservation. The most conserved base-paired stem is P7, which is always 5 base pairs long; moreover, 4 of these base pairs are almost invariant in sequence. The element P6 is remarkable in that in many introns it is only two base pairs long; such a short complementary region would never have been considered significant in the absence of extensive phylogenetic comparison data. Altogether, about 50 nucleotides show some degree of conservation, and about 20 are highly conserved or invariant. These nucleotides must be important either directly for catalysis or indirectly in determining the three dimensional structure of the enzyme. As such, they are a primary target for further genetic and physical analysis. Sequence comparison also points out regions of the molecule that are probably not important for function. For example, the three loops are highly variable, and indeed often contain insertions of very long open reading frame sequences. In addition, a subset of group I introns contains an insertion of 30-50 nucleotides between P3 and P7. This insertion probably forms a surface loop that does not alter the structure of the enzyme.

Mutational Analysis: Most of the genetic analysis of group I intron structure published to date has been concerned with confirming the secondary structure discussed above. I have already mentioned the use of compensatory mutations to demonstrate the importance of the P1 pairing. The same type of experiment has been done for the P3 and P7 regions of the catalytic core (38,39). In the case of P3, the results are simple: mutations in either strand result in loss of activity, and compensatory double mutants restore activity. The situation with P7 more complex. Under conditions of low Mg^{++} concentration, mutants in either strand alone are inactive. The double mutant restores activity but the enzyme is inactive at low temperature. Surprisingly, some of the single mutants actually increase activity above wild type levels under conditions of very high Mg^{++} concentration. These results have been interpreted as suggesting that some major conformational transition involving P7, such as melting and reannealing, is a necessary part of the catalytic cycle.

There is clearly great potential for learning more about the structure of group I introns from more intensive genetic analysis. Two laboratories have reported similar systems for selecting revertants

tants in the intron (28,40). The intron has been placed near the
ginning of a bacterial gene encoding β-galactosidase, in the non-
anslated leader. Expression is only obtained if the intron is
pable of removing itself from the mRNA. Since expression is readily
lected or screened for, revertants of defective mutants can be
tained.

In my laboratory, we have chosen to simply synthesize all possible
tants in all of the conserved nucleotides of the catalytic core.
alysis of these single mutants is beginning to define the most
itical regions of the ribozyme, and to differentiate these regions
cording to function. For example, one cluster of mutants affects
ly guanosine binding. However, we expect that the major use of this
brary of mutants will be the identification of pairs or groups of
teracting nucleotides. The mutants are being constructed in such a
y that double and triple mutants can easily be made by subcloning
striction fragments of plasmid DNA. We are most interested in
nding double mutants that have greater enzyme activity than either
ngle mutant: the simplest interpretation of such mutants is that a
ecific nucleotide-nucleotide interaction has been restored by a
mpensatory change. If enough tertiary interactions can be identified
 this type of genetic approach, it may be possible to develop a
asonable three dimensional model of the structure of the enzyme.
nce this type of analysis yields a series of distance constraints,
 expect that the distance geometry methods developed for the
terpretation of 2D-NMR data will be useful for the physical
terpretation of our genetic data.

emical Probes of Catalysis

anosine Analogues: The RNA enzyme shows great specificity for its
anosine substrate. The nature of the guanosine binding site is of
eat interest, and accordingly a large number of nucleotides and
cleotide analogues have been examined for binding to the Tetrahymena
bozyme (41). Binding has been examined in two ways: either by direct
bstitution for guanosine in the splicing reaction, or by competitive
hibition. In many cases, analogues bind poorly to the guanosine
nding site, but do react once bound; in these cases binding is
amined in a mixed substrate reaction assay.

Kinetic analysis suggests the existence of a specific guanosine binding site. At low guanosine concentration, the rate of the reacti is first order with respect to guanosine concentration; at higher concentrations, the reaction becomes zero order with respect to guanosine. This strongly suggests that a guanosine binding site becomes saturated at high concentrations. The measured Km is approximately 20 uM, both for the intact intron, and for the isolate catalytic domain.

Early experiments showed that guanosine was required for the splicing reaction (1), and could not be substituted by adenosine, uridine or cytidine. The stringent specificity for guanosine is one the most striking characteristics of the splicing reaction, and is characteristic of all group I introns that have been examined. It seems likely that this specificity is determined by interactions wit a number of nucleotides in the enzyme, and that changing the specificity would require changes at several positions, thus acccounting for the extreme phylogenetic conservation of the requirement for guanosine.

Bass and Cech (41) examined a large number of guanosine analogue for activity in the splicing reaction. Modifications at the 5' hydroxyl had no effect on splicing activity: the mono-, di- and triphosphates were all as active as guanosine itself in promoting splicing. However, modifications of the 2' and 3' hydroxyls abolish activity. Since the mechanism of the reaction is assumed to be a nucleophilic attack of the 3' hydroxyl on the phosphate at the 5'- exon-intron junction, it is not surprising that changes at the 3' hydroxyl have drastic effects. Thus, 3'-GMP and 3'-O-methyl guanosin are inactive in splicing. Xyloguanosine, in which the orientation of the 3'-OH is changed, is also inactive. It is somewhat more surprisi that modifications at the 2'-OH also have drastic effects: 2'-dGTP, 2'-O-methylguanosine, and araguanosine (in which the orientation of the 2'-OH is changed) are all inactive. Although the 2'-OH may well involved in some interaction that contributes to binding or helps to orient the sugar correctly, Bass and Cech suggested that it was unlikely that loss of one such interaction would completely abolish splicing activity. It therefore seems that some additional role, suc as an effect on the reactivity of the 3'-OH, may also be involved. When tested for competitive inhibition, 2'dGTP, xyloguanosine and

aguanosine showed no detectable inhibition up to a 500-fold excess
er guanosine.However, in a subsequent study (42), using more highly
rified deoxy and dideoxy-guanosine, these compounds were found to
t as competitive inhibitors, with Ki's of 1.1 mM and 5.4 mM
spectively. Even at concentrations where dG does bind, it does not
act, thus showing directly that the 2' OH plays a role in catalysis
well as a role in binding.

Modifications of the imidazole ring (C-8, N-7) do not affect
licing activity (41). Thus 7-methyguanosine and 8-azidoguanosine
iphosphate are fully active. In addition, we have found that 3-deaza
anosine (the generous gift of Dr. Leroy Townsend, Univ. of
chigan), in which N-3 on the 6 membered ring is replaced with C-H,
fully active. However, modications at O-6, N-1 or the 2-amino group
l affect activity (41). Loss of the carbonyl group (2-aminopurine
bonucleoside), or substitution of the oxygen with sulfur (6-
ioguanosine) lead to partial loss of activity, as does substitution
N-1 (1-methylguanosine). The latter effect is somewhat difficult to
terpret, as it could be due either to loss of an interaction, or to
eric interference due to the bulky methyl group. Methylation of the
amino group also leads to partial loss of activity in the case of N-
methylguanosine, or complete loss of activity for N-2,N-2
methylguanosine. While it is possible that the amino group may be
volved in two hydrogen bonds, steric effects cannot be excluded.
deed, the fact that inosine shows partial activity despite loss of
e amino group suggests that the presence of two methyl groups on the
ino group interferes with binding more than can be explained simply
 loss of two hydrogen bonds. This is perhaps not surprising if one
nsiders that the methyl groups would almost certainly conflict with
y hydrogen bond acceptors close enough to interact with the amino
otons. The Km's measured for both 2-aminopurine and inosine were
out 2.5 mM, and no effect on kcat was observed. This 100-fold
crease in Km corresponds to a ΔΔG of about 2.8 kcal/mole, consistent
th a loss of two hydrogen bonds in each case. Loss of more than two
tential hydrogen bonds (purine ribonucleoside or isoguanosine)
sults in complete loss of activity.

The data reviewed above are consistent with four hydrogen bonds
ing made to the base, and at least two to the sugar. There is at
esent no data bearing on other kinds of possible interactions, such

as base stacking effects. The results of these analogue studies have
revealed the complexity of the binding site, and suggest that it mus
be composed of several nucleotides from the ribozyme. In addition, t
large number of interactions that are apparently made to the guanosi
provides a clear explanation for the specificity of the splicing
reaction for guanosine as opposed to the other nucleosides. It is al
clear that these interactions must serve to orient the guanosine, an
in particular the 3'-OH, correctly relative to the attacked phosphat
and that these interactions must therefore contribute substantially
the mechanism of catalysis.

Additional modifications in the ribose moiety have been explored
by Kay and Inoue (43) with interesting results. They examined the
reactivity of GTP oxidized by periodate to the dialdehyde (referred
as GTP=O). They observed little activity at 30°, but saw substantial
cleavage at $42^{o}C$. Analysis of the products, however, showed that the
residue had not become covalently linked to the beginning of the
intron, in contrast to the control reaction with unmodified GTP.
ATP=O, CTP=O and UTP=O were inactive in this cleavage reaction. Kay
and Inoue suggested the following mechanism for the cleavage reactio
stimulated by oxidized GTP. Addition of water to each aldehyde group
followed by loss of one water generates a cyclic intermediate very
similar in conformation to guanosine. Attack of the 3' hydroxyl occu
normally, cleaving the intron, but the modified guanosine adduct is
labile, and spontaneously decays to release the dialdehyde and the
intron. Alternatively, binding of the dialdehyde may somehow stimula
direct hydrolysis of the exon-intron junction. Although the reaction
with the dialdehyde increases with increasing pH (the reaction with
GTP is pH independent), both the direct hydrolysis mechanism, and th
hydrated intermediate proposal would be expected to be pH dependent,
making it difficult to distinguish between these models at present.

The terminal G of the intron almost certainly binds to the same
site during exon ligation and cyclization as the free G does in the
initial reaction. Kay and Inoue (43) also examined the effect of
modification of the terminal G by periodate oxidation. They found tha
the modified nucleotide could no longer yield cyclized products, but
was still capable of stimulating cleavage at the 5'exon-intron
junction. Aniline treatment of the modified RNA results in loss of tl
terminal oxidized nucleotide. When they compared the ability of free

P to cleave oxidized RNA and the eliminated RNA, they found that the
idized RNA was less reactive. They interpreted this result as being
e to competition of the oxidized G for occupancy of the G binding
cket with free G, preventing it from acting. Unfortunately, GTP=O
s not directly tested for competitive inhibition with GTP.

Kay and Inoue (43) also tested 3'deoxy, 3'amino guanosine and
und that it was inactive in splicing. This is somewhat surprising in
ew of the fact that the amino group should be a much better
cleophile than the 3' hydroxyl of guanosine. Either the 2' hydroxyl
ays an essential role in the nucleophilic attack of the 3' hydroxyl
at cannot occur with a 3' amino group, or the orientation of the 3'
ino group is inappropriate. A role for the 2' hydroxyl in catalysis
s opposed to a role limited to binding) is supported by experiments
th three additional analogues. Reduction of 5'exon-intron RNA with
BH4 results in opening of the ribose ring of the terminal guanosine
tween the 2' and 3' hydroxyls; the resulting RNA is still capable of
rcularization, although at a much slower rate. It is possible that
e reduced rate is a result of the 2' and 3' hydroxyls no longer
ing spatially constrained as they are in intact guanosine. In
ntrast, 9-(1,3-hydroxy-2-propoxy)-methyl guanine, which resembles
e previous derivative except for loss of C-2' and the 2'-hydroxyl,
  inactive. Exon-intron modified in the terminal G with
micarbazide, which also changes the relative orientation of the 2'
d 3' hydroxyls, is also inactive.

se Modification as a structural probe: RNA modification by chemical
agents provides a powerful probe of both structure and catalytic
nction. Several approaches have been explored in Cech's laboratory.
rhaps the most powerful of these methods uses a modification of the
emistry originally developed by Peattie (44) for RNA sequencing.
methyl sulfate methylates adenosine residues at N-1, cytosine at N-
 and guanosine at N-7. In the sequencing protocol, only the modified
 are detected, by reduction with NaBH4 followed by aniline induced
ain cleavage. Inoue and Cech (45) showed that the modified A and C
siduues could be detected as sites of reverse transcriptase pausing
 termination, because these modifications interfere with Watson-
ick base pairing. Thus primer extension on a modified RNA template
n be used to detect these modified bases. These modifications are
rticularly useful in detecting nucleotides involved in secondary

structure, as base-pairing protects these nucleotides from the
modifying reagent. Modification of G at N-7 is not detected by rever
transcription.

A second useful modification is the carbethoxylation of A and G
N-7 by diethylpyrocarbonate (DEPC). The resultant ring opening was
detected in the original sequencing method by aniline induced chain
cleavage; Inoue and Cech showed that this modification could also be
detected by reverse transcription.

Of 30 A residues in regions of the catalytic core not expected t
be base paired, 18 were strongly modified by dimethyl sulfate. In
contrast, of 16 A residues expected to be in base paired stems, only
were strongly modified . This result is in remarkable agreement with
the phylogenetic data, and strongly supports the secondary structure
models derived from this data. This secondary structure is also
supported by the results of nuclease sensitivity experiments (36). T
DEPC modification results are less easily interpreted, since the
factors that would protect N-7 from modification are not well
understood. Some bases modified by DEPC do fall within presumed base
paired stems, although most do not.

Methylation-interference experiments are commonly done to examin
the binding requirements of sequence specific DNA -binding proteins.
The DNA sequence is partially methylated, protein is added, and DNA-
protein complex is isolated. The purified DNA is examined as above f
sites of methylation. Nucleotides whose methylation prevents binding
are always unmodified in DNA isolated from complex. This type of ass
could presumably be extended to modification-interference of catalys
by taking advantage of the cis-acting nature of exon-intron cleavage
Exon-intron could be partially modified (e.g. by methylation), allow
to self-splice, and product molecules isolated. Any nucleotides whos
modification interferes with catalysis will not be modified in the
product RNA, which derives from active molecules. We are currently
exploring the use of this methodology to complement and extend data
derived from the analysis of mutants. A very interesting and simple
extension of these ideas would be to look for suppression of the
effect of a particular modification by a specific mutation, or vice
versa.

ER RIBOZYMES

up II Introns

Group II introns, like group I introns, are defined on the basis
common sequences and secondary structures. All mitochondrial and
oroplast introns appear to fall into one of these two classes.
hough the group II introns have no detectable similarity to group I
rons, some members of this class are also self-splicing (16,17).
e major mechanistic difference between the two classes is that the
up II introns do not use the 3' hydroxyl of free guanosine to
tiate the first transesterification reaction; instead, they make
of the 2' hydroxyl of an internal residue. Thus, the products of
first reaction are free 5' exon, ending in a 3' hydroxyl, and a
nched circular (lariat) form of the intron. The second step
volves attack of the 3' hydroxyl of the 5' exon on the downstream
ron-exon junction, displacing the intron in the form of a stable
iat structure, with concerted exon ligation. The formation of
iat structures during the normal splicing pathway raises the
riguing possibility that there is some relationship between these
rons and nuclear mRNA splicing, which follows a similar pathway.
haps the snRNPS that catalyze nuclear splicing carry out in trans
functions carried out in cis by the enzymatic regions of the group
introns.

The donor group in the first reaction is the 2' hydroxyl of an A
idue that is always present as a bulged A within a conserved base
red stem. Deletion or mutation of this residue blocks splicing. It
not known how the enzyme activates this group for nucleophlic
ack.

The mechanism of splice site selection is just beginning to be
erstood. Normally, free 5' exon is not released, but is held in
ce by the enzyme until the second step is completed. Multiple exon-
ding sites have been identified within the intron by mutational
lysis (46). Mutants in the exon that are defective for splicing can
suppressed by mutants in an exon-binding site.

onuclease P

Ribonuclease P is a ribonucleoprotein complex whose RNA componer is essential for activity (12). It is found in diverse bacterial species (gram-negative and gram-positive), and in mammalian cells. Under conditions of high Mg^{++} concentration, the RNA component alone has catalytic activity (13). Although a partial secondary structure model has been proposed, additional phylogenetic data is required to test the model. The main role of the protein appears to be in electrostatic shielding, ie. it allows the strongly negatively charg enzyme and substrate to approach each other.

The enzyme is required for the maturation of precursor tRNAs. Th products of the hydrolysis reaction are a 3' hydroxyl and 5' phosphate. Although this is similar the group I introns, the attacki nucleophile is apparently hydroxide ion. Marsh and Pace (47) showed that the 3' hydroxyl of the RNA component of RNAse P is not involved in the reaction since the activity of the ribozyme is unaffected by periodate oxidation of the terminal base. This result, in combinatic with the pH dependence of the reaction suggests that free hydroxide the initiating nucleophile in the cleavage reaction, in contrast to the group I introns, where the 3' hydroxyl of guanosine, the 5' exon or the intron is involved in each of three consecutive reactions.

Viroids and Virusoids

Viroids are small circular RNA molecules that are plant pathogen They are transcribed by host enzymes into multimeric linear molecule which self-cleave to generate linear monomers, which are then ligate by host RNA ligase to monomer circles. As in the case of RNAse P and the group I introns, the self-cleaving ability of the RNA was proven by showing that RNA made by in vitro transcription was catalytically active (15). A similar self-cleavage reaction has been observed for virusoids, which are satellite viruses, and for transcripts of Newt satellite DNA sequences (48). In all three cases, the mechanism of cleavage, and the active RNA structure appear to be similar. The reaction products are a 5' hydroxyl, and a 2',3' cyclic phosphate, i contrast to both RNase P, which generates a 3' hydroxyl and 5' phosphate, and the self-splicing introns, which leave a 3'hydroxyl.

The rate of self-cleavage of intact dimer is, in vitro, quite slow. Remarkably, a deletion analysis carried out in order to delimi

critical RNA sequences led to the finding that short transcripts
anking the cleavage site showed greatly enhanced activity (15). The
gion defined by deletion analysis coincides with a compact secondary
tucture, called a hammerhead, that is highly conserved among all
lf-cleaving molecules of this class. The enhanced activity of the
aller molecules is probably a result of competition between
ternative secondary structures. Intact viroids probably form a rod-
ke structure that is largely duplex; in longer molecules, this
active form probably dominates over the active conformation.

Although the hammerhead structure is a remarkably small ribozyme,
lenbeck has shown that an even smaller modified version is active
9). In the hammerhead structure, the small closed loops and the
rge loops can occur at different positions, suggesting that the
ops themselves play no role and need not in fact even be closed
ops. Uhlenbeck therefore synthesized two short RNAs of 25 and 19
cleotides that could potentially anneal together to form a
mmerhead. As expected, the 25-mer was cleaved at the correct site;
e 19-mer could then anneal with a second 25-mer, leading in turn to
s cleavage. Thus the 19-mer acts kinetically as an enzyme, although
e active structure is undoubtedly formed from the 19-mer and the 25-
c together. The small size of this ribozyme, its well characterized
condary structure, and the small number of conserved nucleotides
kes it a particularly attractive target for the sort of saturation
netic analysis of structure that has been discussed above.

NA Cleavage by Lead

Yeast tRNA-phe binds four Mg^{++} ions at specific sites with high
finity; three of these can be displaced by Pb^{++} ions which bind at
sitions close to although not identical with those of the Mg^{++} ions.
e positions of the lead ions were determined in the course of
lving the structure of the tRNA by X-ray diffraction analysis.
ring the preparation and characterization of this particular
omorphous derivative, it was found that at high pH, the lead
ostituted tRNA was cleaved specifically between D-17 and G-18 (50).
the basis of the structural data, it was suggested that the most
kely mechanism for the reaction was abstraction of a proton from the
OH of residue 17 by a hydroxide ion coordinated with the lead ion,
llowed by nucleophilic attack of the $2'-O^-$ on phosphate 18, leading

to chain cleavage with formation of a 2'-3' cyclic phosphate and a 5
hydroxyl. The Pb^{++} ion involved is bound to U-59 and C-60, close to
the site of cleavage in the folded tRNA. Thus the tRNA - Pb^{++} comple
is an RNA metalloenzyme, in which one part of the molecule acts like
the enzyme, and another part acts like the substrate.

Other RNA catalyzed reactions: Several additional candidates for
catalytically active RNA molecules have been reported. A short
transcript from bacteriophage T4 has been reported to be a self-
cleaving RNA (51) . Although several controls against nuclease
contamination were reported, including lack of cleavage of a short
oligo flanking the cleavage site, the activity of RNA produced by T7
transcription has not been reported.

Mitochondrial DNA replication is initiated from an RNA primer.
This RNA is synthesized as a longer precursor, which is cleaved by a
processing enzyme at a specific site to generate the mature form. Th
enzyme has an essential RNA component (52). During the the final
stages of assembly of certain RNA viruses such as poliovirus, one of
the viral coat proteins is cleaved at a specific site. Since it is
difficult to imagine a protease having access to the cleavage site,
which is on the inside of the virion, it has been proposed that the
RNA component of the virus may play some role in this reaction.

The RNA components of the various well known cellular RNP
particles are perhaps the most likely RNAs to have as yet undefined
catalytic roles. Given the similarity of snRNP catalyzed nuclear mRN
splicing to self-splicing of group II introns, it seems quite likely
that the RNA components of some of the snRNPs will play a direct rol
in catalysis. It also seems quite likely that the ribosomal RNAs are
more than a structural backbone for the assembly of the ribosome, an
that they play a direct catalytic role in protein synthesis. Many
ribosomal proteins are not essential for protein synthesis, and have
only an auxiliary role. A direct role for RNA is also supported by t
fact that resistance to some protein synthesis inhibitors results fr
mutations that change the rRNA, not the proteins. The evolutionary
argument may the the strongest, however: it is much easier to
understand the origin of protein synthesis by RNA catalysis than by
protein catalysis. In addition to snRNPs and ribosomes, there are ma
other RNPs where it is possible that the RNA will have a critical

le. For example, it has recently been found that the enzyme telomere
rminal transferase, which is responsible for the elongation of
lomeric DNA, has an essential RNA component.

N-PROTEIN ENZYMES: OLD AND NEW

arch for activity in random sequence RNA chains

An important question in terms of the evolution of catalytic
nction is the probability that a given random sequence will display
particular binding activity or catalytic function. There are 4^{100},
approximately 10^{60}, possible RNA sequences 100 nucleotides long.
e can in principle readily synthesize in the laboratory pools of
ndom 100-mers; a few mg of such random sequence RNA would comprise
me 10^{16} different sequences. One can imagine that on the primitive
rth, a much larger number of random chains might have been
nthesized, perhaps on the order of 10^{30}. In order for life to have
olved, at least one of these 10^{30} molecules must have been a
asonably efficient replicase. In order for this question to be
perimentally approachable now, catalytic molecules must arise at a
equency greater than 1 in 10^{16}.

How could rare catalytically active molecules be detected in the
dst of a large pool of random sequence molecules? We have begun to
rk out the technology required for this experiment. The principle
volved is the repeated amplification and selection of the molecules
interest. In the simplest case of a molecule that can specifically
nd an immobilized ligand, selection by affinity chromatography is
ssible, with an expected enrichment of at least 100 fold per cycle.
e enriched RNA can be amplified as follows. The RNAs are initially
nerated by transcription of a pool of synthetic DNA
igonucleotides, made with defined 5′ and 3′ termini, but with a
ndom mixture of sequences in the middle. The defined 3′ end can be
ed to anneal a primer for cDNA synthesis. The defined 5′ end
quence is used for priming second strand synthesis, using a primer
at includes the T7 RNA polymerase promoter sequence. This double
randed DNA can then be transcribed by T7 RNA polymerase, with a
eld of about 2000 transcripts per template, for an overall
plification of approximately 200 fold per cycle.

This series of steps has been tested on a trial RNA molecule, and 5 sequential rounds of amplification have been achieved. This amplification cycle, in combination with selection by affinity chromatography, should allow the selection of 1 molecule from an initial mixture of 10^{16} molecules in 8 rounds of selection and amplification. This procedure could be modified to allow for the selection of catalytically active molecules in several ways. One possibility would be to simply select for binding to transition state analogues; some of these might catalyze the corresponding reaction, as has been found for antibodies raised against transition state analogues.

Prochiral Nucleotides

Joyce et al. (53) have noted a number of serious problems with the idea that the original self-replicating genetic system could have been based on RNA. Prominent among these difficulties are the lack of an efficient prebiotic synthesis for ribose, the complex mixtures of sugars generated in the formose reaction, the instability of the sugars, the inefficiency of the synthesis of purine β-ribosides, the lack of any mechanism for the synthesis of pyrimidine nucleosides, and potentially worst of all, the inevitable prebiotic synthesis of a racemic mixture of stereoisomers of nucleotides. Joyce et al. noted that the spontaneous polymerization of activated nucleotides of one enantiomer on a template strand would be inhibited by the presence of the other enantiomer. They referred to this phenomenon as enantiomer cross inhibition, and demonstrated its occurence with a chemical model system, the polymerization of guanosine 5'-phospho-2-methylimidazole on a poly-C template. Polymerization of the D-isomer is efficient; while incorporation of the L isomer occurs, it is less efficient, and once incorporated it acts as a chain terminating nucleotide. Therefore, far from a template of one specificity acting to select subunits of like specificity from a racemic mixture, the whole autocatalytic process of self replication is poisoned by the existence of a mixture of enantiomers.

In order to circumvent this problem, Joyce et al propose that simpler flexible or even prochiral nucleotides formed the subunits of the original self-replicating polynucleotides. They discuss several candidate acyclo-nucleotides, all of which are much more flexible in

nformation than the closed ring ribose based nucleotides. Such
yclonucleotides may also be easier to synthesize prebiotically. The
exibility of the sugar portion means that the double helical
ructure can be retained, while problems of chain termination are
oided. It is even possible that heterogeneity of the backbone could
 tolerated. A primitive self-replicating system may have arisen
sed on such nucleotides, and only later have been replaced by a more
ficient and homogeneous RNA based replicase.

rbohydrates

The case for a simpler progenitor of RNA may be taken even further
an in the above discussion. Polymers of glyceric acid (and possibly
her simple sugars) may be capable of forming defined secondary and
rtiary structures, and hence might be capable of performing
talytic functions (54). This possibility arises from the observation
at homopolymers of L-glycerate or D-glycerate become insoluble at
latively low molecular weights, whereas heteropolymers of the two
ereoisomers that are of much higher molecular weight remain soluble.
e insolubility of the homopolymers may arise from the formation of
crocrystalline domains formed by hydrogen bonding between
tiparallel strands. In heteropolymers, the size of such domains
uld be limited by the interspersion of the L and D isomers, and
mplex structures could potentially arise. While the complexity of
cromolecular carbohydrate structures is well documented, no case of
catalytically active polysaccharide has yet been found.
vertheless, the ready prebiotic synthesis of sugars and sugar
osphates makes the idea of a prebiotic role for complex
lysaccharides in the origin of life attractive. The recent
servation of a double helical structure for a short segment of a
arch molecule (actually a modified malto hexosaccharide) is
rticularly intriguing. Perhaps nucleotides are best considered as
dified sugars, with the sugar portion of the molecule predating the
re complex modified derivatives.

vel Heteropolymers

An interesting area for future research is the synthesis of
dered heteropolymers, analagous to nucleic acids and proteins, but
mposed of novel monomeric subunits. The choice of appropriate

monomers could facilitate the design of heteropolymers with a
predictable three-dimensional structure. The prediction of tertiary
protein structure from primary sequence is a difficult problem that
will probably remain unsolved for some time, simply because of the
large number of possible interactions between the 20 different amino
acid subunits. The multiplicity of these interactions makes even the
prediction of secondary structures difficult. On the other hand, the
peptide backbone is quite simple, with only two dihedral angles per
repeat, and places substantial constraints on the available
conformations. Nucleic acids have the opposite advantages and
disadvantages: with only four types of monomer units, and very
specific nucleotide-nucleotide interactions, the prediction of
potential secondary structures is trivial, and the main problem is
distiguishing between the many possible competing potential
structures. In addition, the backbone is so flexible that it places
few constraints on allowed structures. Finally, the number of hydrog
bond donors and acceptors on each nucleotide is large enough that th
number of potential base pairs or base triples is rather unwieldy. O
might therefore imagine that the ideal set of monomers, in terms of
yielding polymers of predictable structure, would have the following
characteristics. The number of types of monomers should be 6 or 8, f
fewer than the number of amino acids, but large enough to result in
unambiguous secondary structures. These monomers should be able to
form 3 or 4 base pairs analagous to Watson Crick base pairs, for the
formation of helical regions, and a strictly limited set of addition
possible interactions to allow for tertiary structures. The synthesi
of ribo- and deoxyribo-nucleotides capable of forming new 'Watson-
Crick'-like base pairs is under way in the laboratory of Steven A.
Benner. The chemical backbone of this hypothetical polymer should be
simple, and more restrictive than that of nucleic acids. Of course,
the constraints that are imposed with the view of facilitating
structural prediction must not be so severe that they prevent the
formation of any intereseting structures at all! These subunits woul
have to be available in suitable protected forms for solid phase
assembly into ordered heteropolymers, since existing enzymes will no
use them as substrates for polymerization. Both the subunits and the
assembled backbone repeats must be stable to chemical and enzymatic
degradation, at least if the resulting polymers are to have any
application as industrially usable catalysts.

lf-replicating polymers

Life may be said to have begun with the emergence of self-
plicating polymers - presumably some type of nucleic acid, if not
A itself. Current research in my laboratory is oriented towards
derstanding the mechanism by which group I intron RNA molecules are
pable of catalyzing transesterification reactions. This reaction is
 particular interest to us because it is essentially the same
action that RNA polymerases, or any self-replicating RNA, must
talyze. Indeed, one very simple strategy for a replicase is merely
 splice together a series of short oligonucleotides on a longer
mplate strand. In the extreme, this reduces to splicing together
nucleotides, with the release of one monomer for every monomer
corporated into the growing chain. In terms of catalysis, this is
ry close to the polymerization of activated nucleotide monomers,
ether the monomers are activated by a triphosphate as in living
lls, an imidazolide group, as in the model systems studied by Orgel
d others, or any other suitable leaving group. Modification of the
bstrate specificity of the group I intron to allow the
lymerization of monomers on a template may be possible. The exon
gation reaction is essentially the joining of two oligos on an
nternal' template. Cech's lab has extended this to the
lymerization of C residues on the same internal template, as
scussed above. If the enzyme can be persuaded to use an external
mplate, in a reaction analagous to trans-splicing, and to relax the
cleotide specificity that has presumably evolved to ensure splice-
te specificity, template directed polymerization should be possible.
 the template is another copy of the enzyme sequence (or its
mplementary sequence), then autocatalytic self-replication should
sue. Even if the original molecule is very inefficient, selective
essures should force the evolution of more efficient replicases in
tro, at least if the RNA molecules can be compartmentalized, perhaps
 vesicles, so that a given molecule copies its own progeny and not
related sequences. Of course, a self-replicating polymer in a
mipermeable vesicle would constitute a simple cell, and the creation
 such structure in the laboratory would open up many new areas of
search.

REFERENCES

1 Cech TR, Zaug AJ, Grabowski PJ (1981) Cell 27:487
2 Kruger K, Grabowski PJ, Zaug AJ, Sands J, Gottschling DE, Cech TR (1982) Cell 31:147
3 Woese CR (1967) The Origins of the Genetic Code, Harper and Row
4 Crick F (1968) J Mol Biol 38:367
5 Orgel LE (1968)J Mol Biol 38:381
6 Visser CM, Kellog RM (1978) J Mol Evol 11:171
7 Gilbert W (1986) Nature 319:618
8 Benner SA, Ellington AD (1987) Nature 329:295
9 Weiner AM, Maizels N (1987) Proc Natl Acad Sci USA 84:7383
10 Chamberlin M, Ryan T (1982) In: Boyer P (ed) The Enzymes p84, Academic Press
11 Melton DA, Krieg PA, Rebagliati MR, Maniatis T, Zinn K, Green MR (1984) Nuc Acids Res 12:7035
12 Stark BC, Kole R, Bowman EJ, Altman S (1977) Proc Natl Acad Sci U 75:3719
13 Guerrier-Takada C, Gardiner K, Marsh T, Pace N, Altman S (1983) Cell 35: 849
14 Prody GA, Bakos JT, Buzayan JM, Schneider IR, Bruening G (1986) Science 231: 1577
15 Hutchins CJ, Rathjen PD, Forster AC, Symons RH (1986) Nuc Acids R 14:3627
16 Peebles CL, Perlman PS, Mecklenburg KL, Petrillo ML, Tabor HJ, Jarrell KA, Cheng HL (1986) Cell 44:213
17 Van der Veen R, Arnberg AC, Van der Horst G, Bonen L, Tabak HF, Grivell LA (1986) Cell 44:225
18 Chu FK, Maley GF, West DK, Belfort M, Maley F (1986) Cell 45:157
19 Bonnard G, Michel F, Weil JH, Steinmetz A (1984) Mol Gen Genet 194:330
20 Tabak HF, Van der Horst G, Osinga KA, Arnberg AC (1984) Cell 39-6
21 Kan NC, Gall JG (1982) Nuc Acids Res 10:2809
22 Garriga G, Lambowitz AM (1984) Cell 38:631
23 Akins RA, Lambowitz AM (1987) Cell 50:331
24 Davies RW, Waring RB, Ray JA, Brown TA, Scazzocchio C (1982) Natu 300:719
25 Szostak JW (1986) Nature 322:83
26 Zaug AJ, Kent JR, Cech TR (1984) Science 224:574
27 Inoue T, Sullivan FX, Cech TR (1986) J Mol Biol 189:143
28 Waring RB, Towner P, Minter SJ, Davies RW (1986) Nature 321:133
29 Been MD, Cech TR (1986) Cell 47:207
30 Been MD, Cech TR (1985) Nuc Acids Res 13:8389
31 Zaug AJ, Been MD, Cech TR (1986) Nature 324:429
32 Zaug AJ, Cech TR (1986) Science 231:470
33 Cech TR (1986) Proc Natl Acad Sci USA 83:4360
34 Zaug AJ, Cech TR (1986) Biochemistry 25:4478
35 Michel F, Dujon B (1983) EMBO J 2:33
36 Cech TR, Tanner NK, Tinoco I, Weir BR, Zuker M, Perlman PS (1983 Proc Natl Acad Sci USA 80:3903
37 Burke JM, Belfort M, Cech TR, Davies RW, Schweyen RJ, Shub DA, Szostak JW, Tabak HF (1987) Nuc Acids Res 15:7217
38 Burke JM, Irvine KD, Kaneko KJ, Kerker BJ, Oettgen AB, Tierney WM Williamson CL, Zaug AJ, Cech TR (1986) Cell 45:167
39 Williamson CL, Tierney WM, Kerker BJ, Burke JM (1987) J Biol Chem in press
40 Price JV, Cech TR (1985) Science 228:719
41 Bass BL, Cech TR (1984) Nature 308: 820
42 Bass BL, Cech TR (1986) Biochemistry 25:4473
43 Kay PS, Inoue T (1987) Nuc Acids Res 4: 1559
44 Peattie DA (1979) Proc Natl Acad Sci USA 76:1760

Inoue T, Cech TR (1985) Proc Natl Acad Sci USA 82:648
Jacquier A, Michel F (1987) Cell 50:17
Marsh TL, Pace NR (1985) Science 229:79
Epstein LM, Gall JG (1987) Cell 48:535
Uhlenbeck OC (1987) Nature 328:596
Brown RS, Dewan JC, Klug A (1985) Biochemistry 24:4785
Watson N, Gurevitz M, Ford J, Apirion D (1984) J Mol Biol 172:301
Chang DD, Clayton DA (1987) Science 235:1178
Joyce GF, Schwartz AW, Miller SL, Orgel LE (1987) Proc Natl Acad
 USA 84:4398
Weber AL (1987) J Mol Evol 25:191

RECONSTRUCTING THE EVOLUTION OF PROTEINS

Steven A. Benner

Laboratory for Organic Chemistry
E.T.H. Zurich, CH-8092 Switzerland

Abstract: Data from biological chemistry, including protein structure, enzymatic mechanisms, and metabolic pathways, can be unified by understanding the distinction between selected and non-selected behaviors. We outline methods for constructing functional and historical models explaining these behaviors, and show how they can be applied to organizing biochemical data, tested experimentally, and used to engineer the behavior of proteins using recombinant DNA techniques.

INTRODUCTION

To the organic chemist, it is never sufficient simply to "explain" the reactivity of organic molecules. Rather, the chemist must go one step further, to "control" the behavior of molecules by altering their structure in a controlled way. This is, in fact, a rather stringent definition of "understanding," as it requires the "prediction" of behavior from structure (or structure from behavior).

Biological chemistry presents a special problem to the organic chemist. Although the tools for synthesis, purification, and structural characterization are now available for manipulating rather large biological macromolecules (proteins and nucleic acids in particular), the theory supporting these manipulations is inadequate. We do not know enough to control the behavior of biological macromolecules; still worse, it is not clear that we know enough even to design synthetic molecules to expand our understanding about how reactivity in such organic molecules might be controlled.(1) Starting from scratch, there are simply too many oligopeptides to make; starting from native proteins, there are simply too many structural alterations (mutations) to introduce.

This symposium spans the range chemistry to molecular biology. However, each speaker has developed approaches for solving the problems inherent in complexity of biological macromolecules.

S. A. Benner (Ed.)
Redesigning the Molecules of Life
© Springer-Verlag Berlin Heidelberg 1988

Richard Kellogg has abstracted the details of biochemical
reactivity, and rebuilt these into designed organic "model
systems." Alan Fersht has shown how a sufficiently large
collection of structural variations in a natural enzyme can, if
supplemented with rigorous physical organic analysis, yield a
sophisticated picture of the origin of catalysis. Tom Kaiser has
used a chemical understanding to bridge the gap between structure
and biological activity of small polypeptides, using abstractions
of natural polypeptides to design synthetic polypeptides that
mimic their biological activity. Finally, Jack Szostak is
showing how a molecular biological approach enables the scientist
to study and perhaps recreate catalytic activity in RNA
molecules.

Today, I shall develop a different but related theme. We begin
by observing that nature gives us, at low cost, a collection of
proteins that is remarkably diverse, both structurally and
behaviorally. Determining the primary sequences of these
proteins and studying the intricacies of the behaviors of these
proteins are both growing industries. Some of the behaviors of
these enzymes are highly engineered by evolution; others are not
obviously so. What information can we extract from this
diversity, what does it mean in terms of the proteins' past
histories, how can it be interpreted in relation to biological
function, and how can we use it to "engineer" the behavior of
proteins?

Native proteins are the product of evolution, not design, and
evolution itself is a combination of two quite different
processes, natural selection and neutral drift.(2,3) Natural
selection guides the evolution of enzymatic behaviors that
influence the ability of a host organism to survive and
reproduce. These behaviors can be interpreted in terms of
biological function, and their evolution can be described as a
"search" for optimal behaviors. Neutral drift describes the
evolution of traits of enzymes that make no difference to the
survival of the host. Drift randomizes behavior in proteins, and
will destroy behaviors that are potentially functional if they
are not constrained from drift by natural selection. Indeed,
selection and drift are in many ways opposite in their effect.
Drifting behaviors cannot be interpreted in terms of biological
function, and research efforts that focus on them need not offer

any information about how to control the behavior of
proteins.(4)

The separation of enzymatic behaviors into these two classes is
complicated by the possibility that traits that are not directly
functional might be structurally coupled to traits that are. For
example, the choice of a particular stereospecificity (retention
or inversion) in an enzymatic reaction may not directly influence
the survival of a host organism; however, once chosen, it may be
difficult to evolve a protein with the opposite stereospecificity
because, to do so, structure at the active site must be disrupted
so much that behaviors (e.g., catalytic power) that are selected
are also disrupted. Although such "non-functional-but-coupled"
traits are not optimized and do not reflect function, neither
they they drift. These traits reflect randomly conserved
historical accidents in the history of the protein, and the
peculiar nature of the structural link between them and the
functional trait that constrains them from drifting.

We cannot interpret the diversity in structure and behavior in
proteins presented to us by nature, much less use this diversity
as a starting point for engineering the behavior of proteins,
unless we distinguish between selected and non-selected traits.
Making this distinction is difficult, and requires many
techniques. We shall emphasize three, logical model building,
inspection of natural proteins, and deliberate design of
proteins. Below are summarized general conclusions about the
relationship between structure, behavior, and selection in
proteins, and data supporting these conclusions.

SELECTION, DRIFT, AND CONSERVATION

Structure And Behavior
We can say the following things about the relationship between
structure and behavior in enzymes.

A. Point mutations can create substantial variation in behavior:
"Substantial" refers to a perception of the bio-organic chemist
examining the behavior in vitro. For example, single amino acid
substitutions in ribosomal proteins raise and lower the fidelity
of translation.(5) A single base change is sufficient to destroy
the promoter activity in certain genes. (6) Six mutations in
alcohol dehydrogenase from horse liver significantly alter the
substrate specificity of the enzyme. (7) A single amino acid

substitution in 5-enolpyruvylshiimate-3-phosphate synthase alters
the ability of the enzyme to be inhibited by the herbicide
glyphosate.(8)   A single amino acid change alters antigenicity
and replication in an influenza virus.(9) Point mutations
introduced by recombinant DNA technology have been found that
completely destroy catalytic acitvity, increase or decrease
stability of the protein, alter regulatory properties, substrate
specificity, or biological function.(10)

B. However, most point mutations have little impact on behavior:
For example, Miller and coworkers collected nearly one hundred
nonsense mutations (those introducing a stop signal into the
gene) in the gene coding for the lac repressor protein.(11) The
majority of these had no detectable impact on behavior. More
recently, Fersht and his colleagues have introduced many dozen
mutations into the active site of tyrosyl aminoacylt-RNA
synthetase. Even though these mutations were virtually all at
the active site, the the impact on behavior of many could be
characterized as "small".(12)

C. Even in cases where a point mutation causes a substantial
change in behavior, the behavioral change generally can be
"suppressed" by alteration at another site:
This has been demonstrated for catalytic activity and stability,
and should apply to other behaviors. For example, some random
mutations in staphylococcal nuclease lowered the stability of the
folded form of the protein. (13,14) The effects of these
mutations on enzymic stability could be suppressed by mutations
at a second site. Several of these "second site suppressors"
were "global" i.e., they could improve the stability of the
protein damaged by mutations at many other sites. Interestingly,
one of these global suppressors is present in a naturally
occurring staphylococcal nuclease.(15)

Analogously, replacement of the catalytically important glutamate
by aspartate in triose phosphate isomerase produces a mutant
enzyme 10,000 fold poorer as a catalyst. Mutation at a second
site restored a large fraction (100 x) of the catalytic
activity.(16,17)

D. The relation between behaviors and structure can be scaled:
Examination of the divergence in behaviors of homologous proteins
from nature suggests that alteration of some behaviors requires
fewer structural alterations than other. This is a measure of the

"linkage" between functional and non-functional behaviors; it does not necessarily imply that the highly conserved behaviors are functional. Thus, kinetic behavior drifts/adapts faster than quaternary structure, quaternary structure faster than substrate specificty, substrate specificity faster than stereospecificity, stereospecificity faster than catalytic mechanism, and catalytic mechanism faster than gross tertiary structure (Table 1).(10)

Two examples illustrate how such scales can be constructed. Pancreatic and seminal RNAs have identical amino acids in 81% of the positions.(18) Yet their quaternary strutures are different (one is a dimer, the other a monomer),(19) their substrate specificities are different (one acts on single stranded nucleic acid, the other prefers double stranded nucleic acid),(20) and their biological activities are different (seminal RNase has potent antitumor activity, pancreatic RNase has no antitumor activity). (21,22) Other behaviors diverge only after greater structural divergence. "Angiogenins," proteins thought to have a role in the vascularization of solid tumors, have sequences that are only 40% identical to bovine pancreatic RNase.(23,24) At this level of sequence divergence, the number of disulfide bonds has diverged; angiogenins have 3, mammalian pancreatic RNAses have 4.

Alcohol dehydrogenases show greater divergence in behavior. The 3 isozymes from yeast (95% identical) of alcohol dehydrogenases have different substrate specificities and stabilities.(25) The enzymes from yeast and horse liver (40% identity) have grossly different substrate specificities, kinetic properties, and quaternary structures.(26) Glucose dehydrogenases and ribitol dehydrogenases (25% identity)(27) catalyze somewhat different types of redox reactions. Glucose dehydrogenase catalyzes the oxidation of a hemiacetal, while ribitol dehydrogenase catalyzes the oxidation of a simple alcohol. The alcohol dehydrogenases from Drosophila and yeast (with 25-30% sequence similarities in one domain only) catalyze the same reaction, but with different mechanisms (metal versus no metal), substrate stereospecificities, and cofactor specificities.(4)

Of course, the numbers in Table 1 give an upper limit to the amount of sequence divergence that is needed to produce a specific divergence in behavior. For example, trypsinogen and chymotrypsinogen have sequences that are 39% identical. They

Table 1
Specific Examples Relating Behavioral and Structural Divergence

<u>Variable</u> <u>by</u> <u>Point</u> <u>Mutation</u>
Kinetic properties: k_{cat}, K_M, internal equilibrium constants
Regulatory properties, allosteric inhibition and activation
Thermal stability, substrate specificity, solubility, biological
Stereospecificity
    Cofactor stereospecificity in yeast ethanol dehydrogenase
<u>Variable</u> <u>with</u> <u>10%</u> <u>sequence</u> <u>divergence</u>
Substrate specificity
    Ethanol vs sterols (liver alcohol dehydrogenase) (2%)
<u>Variable</u> <u>with</u> <u>20%</u> <u>sequence</u> <u>divergence</u>
Quaternary structure
    Seminal (dimer) and pacreatic (monomer) ribonuclease (19%)
<u>Variable</u> <u>with</u> <u>30%</u> <u>sequence</u> <u>divergence</u>
Number of introns in gene
    Lysozyme
<u>Variable</u> <u>with</u> <u>40%</u> <u>sequence</u> <u>divergence</u>
Substrate specificity (151)
    2-Oxoglutarate dehydrogenase vs pyruvate dehydrogenase 45%
<u>Variable</u> <u>with</u> <u>50%</u> <u>sequence</u> <u>divergence</u>
Substrate specificity (152-154)
    cycloisomerase 1 vs. clc B muconate cycloisomerase
    tryptophan hydroxylase vs. phenylalanine hydroxylase (49%)
    tyrosine hydroxylase vs. tryptophan hydroxylase (55%)
    phenylalanine hydroxylase vs. tyrosine hydroxylase (55%)
    acetylcholinesterase vs butyrylcholinesterase (45%)
Mechanistic differences (155))
    Superoxide dismutases (Mn vs. Fe) 49%
Reaction type (156)
    Phosphoglycerate mutase vs. diphosphoglycerate mutase (49%)
<u>Variable</u> <u>with</u> <u>60%</u> <u>sequence</u> <u>divergence</u>
Intradomain disulfide bonds
    Mammalian RNase (4) vs. turtle RNase and angiogenin (3) (60%
Reaction type (157)
    Tryptophan synthetase vs. threonine synthetase (58%)
<u>Variable</u> <u>with</u> <u>70%</u> <u>sequence</u> <u>divergence</u>
Reaction types (158)
    Eukaryotic repressor qa-1S (Neurospora crassa) vs.
        Shikimate dehydrogenase (S. cerevisiae) 26%
<u>Variable</u> <u>with</u> <u>80%</u> <u>sequence</u> <u>divergence</u>
Reaction type
    Fumarase vs. aspartase (76%)
<u>Variable</u> <u>with</u> $\geq$ <u>80%</u> <u>sequence</u> <u>divergence</u>
Mechanistic differences and stereospecificity (159)
    Alcohol dehydrogenases (Zn^{++} vs. no Zn^{++})
"Essential" active site residues diverge
    Lysozymes (homology by x-ray strcture)
Reaction type (160,161)
    Malate synthase (cucumber)/Uricase (soybean
        nodule)/glycolate oxidase (spinach) (marginal alignmen
    Fructose-2,6-bisphosphatase vs. phosphoglycerate mutase
        (active site peptides
Different binding types (246)
    Corticosteroid binding globin vs. serine protease inhibitors

have different substrate specificities. Thus, fewer than 61% of
the residues must be changed to change substrate specificity.
Exactly how much of the 61% difference is necessary for the
difference in substrate specificity is not known. It has been
argued that as few as a single change in sequence could produce
the same effect. This argument is now being directly tested
using site specific mutagenesis.(28)

E. Conclusions::
The principal concern in the adaptation of modern proteins is
whether the behaviors we observe reflect "global optima," or
whether behaviors are "trapped" in local optima because
structures with optimal behaviors are isolated from starting
structures by structures with unacceptable behaviors. The data
summarized above suggests the former. The existence of mutations
that create substantial variation in behavior suggests that
substantially altered behavior is never more than a few
mutational steps away from any starting structure. The existence
of many behaviorally silent mutations, together with suppressor
mutations suggests that proteins with improved behaviors are not
isolated from starting structures by proteins with unacceptable
structures.

Structure And Selection
We can draw the following conclusions about the relationship
between structure and selection in proteins.
A. Many point mutations influencing structure only slightly are
nevertheless selectable:
Kreitman has recently shown that natural selection can control
molecular structure of proteins down to a single amino acid. In
the wild, the gene for the alcohol dehydrogenase from Drosophila
(29) can be found in many different structural forms. Given the
level of structural polymorphism in the gene, one would expect
several dozen variants of the protein in the wild. In fact, one
sees only two, and these variants correlate with latitutde and
altitude. The remainder of the structural variation at the
genetic level is "silent"; it occurs in the third base of codons
or in non-functional parts of introns. The inescapable
conclusion is that structural variations in the gene that were
not silent, but led to altered proteins, were sufficiently
disadvantageous that they were removed by natural selection.

B. However, such stringent selection does not apply to all

macromolecular structures:

There are certain genetic structures that are not functionally
constrained from drifting. For example, pseudogenes are DNA
sequences that resemble sequences for catalytically active
proteins but which cannot themselves code for an active protein.
(30) Often, pseudogenes are truncated versions of active genes,
or contain stop codons internal to the coding sequence, and
therefore would produce only fragments of proteins if they were
translated. Pseudogenes are believed to arise from a duplication
of an active gene.(31)

The structure of pseudogenes drifts rapidly (on the order of 10^{-9}
to 10^{-8} changes/site/year).(31-34) Similarly, codon usage in
higher organisms also drifts rapidly. Aside from confirming the
notion that these structures are non-functional, these
quantitative measures of the rate of drift of non-functional
behaviors can be used to assess function generally. If the rate
of of structural divergence in a protein is slower, it indicates
functional constraints on drift.

For example, the detailed structure of albumins is believed not
to reflect function; the proteins apprear to serve no selected
catalytic role, and genetically transmitted variations in the
structure of albumin in humans are not associated with clinical
symptoms.(35) Consistent with this belief, the rate of
divergence of structure in albumins is on the order of 5-7 x 10^{-9}
changes/site/year, comparable to the rate of drift of
pseudogenes. The implication is that wide variation in the
structure of albumin has little impact on the fitness of the host
organism. Likewise, the structure of the C peptide of pre-
proinsulin (the portion that is removed proteolytically to create
an active insulin) diverges at a rate of 7 x 10^{-9}/site/year.(36)
The rate of change of fibrinopeptides is 6 x 10^{-9}/site/year.(37); The rate of amino acid substitutions of the
alphafetoprotein gene is 1.5 x 10^{-9}/site/year. (38). In each
case, the rapid rate of drift is consistent with few functional
constraints on the structure of the protein. This is not
surprising, for in each case, the peptide is discarded in the
process of creating a functional protein.

Natural variation in the structure of a single enzyme within a
population (polymorphism) also suggests that at least some
stuctural variation is neutral.(39) For example, an esterase in

Drosophila displays considerable structural polymorphism at the protein level. In man, many naturally occurring variants of hemoglobin are not associated with disease.(40)

C.In a few cases, evidence for positive adaptation are now known: Positive adaptation refers to the introduction of mutations at a rate faster than that of neutral drift, indicative of a protein rapidly evolving to keep up with a changing environment. For example, both growth factors (41) and protease inhibitors (42) have structures that are diverging faster than those of pseudogenes, suggesting that the structural variation is directly selected in a rapidly changing environment.

In a very recent work, an example of convergent evolution of primary structure (as opposed to well-known examples of convergent evolution of behavior and tertiary structure) has been observed in lysozymes from ruminants.(43) The sequence of lysozyme from a recently evolved branch of ruminant primates is more similar to the sequence of lysozyme from ruminant artiodactyls (e.g., cow) than it is to that of lysozymes from non-ruminant primates. As digestive lysozyme in ruminants serves a special role in digesting the walls of bacteria growing in the rumen, the discovery of sequence convergence in lysozymes from divergent lines of mammals is strongly suggestive of functional adaptation.

D. Scales of selectability can be constructed:
The degree to which a particular biological macromolecule is under selective pressure depends on its function, its position in the heirarchy of information flow, and the organism that it is in. Here again, scales can be constructed.

1. The impact of a structural change on survival value often follows simple chemical intuition. Conservative substitutions (e.g., replacing lysine by arginine) appear more likely to be approximately "neutral" than non-conservative substitutions. Altering residues on the surface of a protein has, in general, less impact on survival value than altering residues inside a protein. Surface residues drift faster than residues interior to the protein. "Conservative" substitutions are observed more frequently than "non-conservative" substitutions, even after correcting for the degeneracy of the genetic code.(44)

2. A second scale relates impact on survival to the position in

the heirarchy of information transfer where the structural perturbation occurs. Thus, including unused DNA in a genome incurs a cost associated with the replication of the DNA at each generation.(45) This cost is ca. 2 ATP equivalents per excess base per generation, plus the cost required to synthesize the additional bases, plus a cost associated with maintenance and repair.

More expensive is the undesired deregulation of the gene leading to the synthesis of unused protein. In a "typical" prokaryote, a gene can make 10^3 copies of a protein per minute, requiring 4 ATP units for each amino acid (including the cost of the message) in each protein expressed, the ATP's needed to synthesize the amino acids, and additional overhead costs. Thus, expressing unused protein in E. coli is roughly 10^6 times more expensive per gene than carrying excess DNA.

Most expensive is the undesired catalytic activity of an expressed protein, provided that the protein catalyzes a reaction that is not physiologically at equilibrium. As enzymes dissipate free energy by allowing a system coming to chemical equilibrium, the cost of an enzyme doing undesired reactions is the cost of undoing those reactions. Thus, a protease can cut 10^3 peptide bonds per minute, requiring the expenditure of roughly 10^4 ATP molecules to repair them. Thus, the catalytic activity of an undesirably active protein costs roughly 10^{12} times more than carrying the DNA silently in the chromosome.

Such a scale is qualitatively entirely consistent with what is known about genetic regulation and evolution in microorganisms. However, it is important to note that measuring quantitatively the "cost" associated with the waste of energy in synthesizing an unused protein remains an unsolved challenge. (46-49)

3. The impact on survival of a mutation depends on the size and complexity of the host organism, and is largest in viruses, then prokaryotes, then single cell eukaryotes, and smallest in multicellular eukaryotes. A variation in behavior of a specific magnitude in a bacterium represents a larger fraction of the "total behavior" that it does in a mammal. Further, much of the survival of a multicellular animal presumably depends on physiology, a factor influencing the survival of bacteria far less.

4. The extent to which structural variation escapes the attention of natural selection depends on the chemical process in which the protein is involved. For example, because the geometric requirements for binding are chemically less stringent than for catalysis, binding proteins should display greater non-selected structural variation than catalytic proteins.

Further, we often can qualitatively estimate the "difficulty" of catalyzing particular reactions, using either organic chemists intuition or less subjective measures such as the stereoelectronic demands on a reaction.(50) For example, the antibonding orbital accepting electron density from an attacking nucleophile is larger in an esterase than in an amidase. Based on this observation, structural variation in esterases is expected to influence rate of reaction (and hence is more likely to be neutral) than structural variation in amidases. Likewise, redox reactions involving electron transfer (e.g., perhaps xanthine oxidase) are geometrically more demanding than reactions involving hydride transfer (e.g., alcohol dehydrogenase). The more abundant (and apparently non-selected) polymorphism in xanthine oxidase from Drosophila than in corresponding alcohol dehydrogenases are consistent with, but are far from proof of, this notion.(39)

The value of scales is that they permit the construction of a fortiriori arguments about the selectability of behaviors for which we may have no specific data. Thus, if single point mutations in alcohol dehydrogenase from Drosophila are the result of natural selection, it is plausible to argue that the multiple forms of alcohol dehydrogenase in yeast are also.(25)

A good illustration of the application of these scales is the issue of codon selection, a structural trait that might either be functional or non-functional. Weissmann and his coworkers have shown that variations in codon usage influence the relative numbers of Q-beta virus in populations;(51) the implication is that in viruses, codon selection reflects selective pressures. In bacteriophage T7, bias in codon use also suggests that codons choice is a selected trait. (52)

In bacteria and yeast, analysis of codon usage suggests that codon selection correlates with tRNA abundance in E. coli and in yeast.(54-57) Again, the variation in codon usage is presumed to be functional, perhaps controlling the level of gene expression. However, codon usage does not appear to be selected in higher organisms. The rate of drift in codon use is only slightly slower than the rate of drift in the structure of pseudogenes or introns.(31) Although there have been many suggestions that the selection of codons in higher organisms is biased.(58-61) the data are statistically marginal, and in many cases, no bias at all can be detected. (62)

Selection And Behavior
We can say the following things about the relationship between behavior and selection.

A. In behaviors that can be altered by changing only a few amino acids, patterns of behavior consistent with functional expectations must reflect functional adaptation:

This fact is best illustrated by kinetic parameters, which (vide supra) can be greatly altered by single amino acid substitutions. This means that if that identical kinetic behaviors are observed in divergent enzymes, they must reflect functional constraints on drift that act directly on kinetic parameters, rather than on some other behavior that is structurally coupled to kinetic behavior. The quantitative kinetic behavior of enzymes can be used to set an upper limit on the extent of behavioral variation that is selectively neutral. For example, the k_{cat}/K_M values for certain enzymes varies with the second order diffusion rate constant in physiological media within a factor of 10 in widely divergent enzymes.(10,63) Michaelis constants generally are within an order of magnitude of physiological substrate concentrations; again, this observation holds over a range of widely divergent enzymatic structures. As k_{cat} and K_M could drift with small changes in primary structure were they not functionally constrained, this is a clear statement that kinetic behavior is constrained from drifting to within a factor of 10 (corresponding to 1.4 kcal/mol).

Recent evidence suggests that natural selection controls behavior much more precisely. For example, the internal equilibrium constants (reflecting the free energy difference between enzyme-bound substrates and enzyme-bound products) of an enzyme is expected on theoretical grounds to be "downhill" in the direction of metabolic flux in an enzyme, as this arrangement provides the enzyme with the greatest catalytic efficiency under a particular

set of physiological conditions.(64) This notion has been
tested using the isozymes of lactate dehydrogenases, one from
muscle (where the flux is in the direction of lactate), and one
from heart (where the flux is in the direction of pyruvate). The
internal equilibirum constants of the two lactate dehydrogenases
are different in the direction predicted by theory, and appear to
be "tuned" to within 0.4 kcal/mol.(65)

One might even conclude that these behaviors are optimal (as
opposed to simply constrained from drift). As site directed
mutagenesis studies suggest that proteins with increased
catalytic power are accessible, (66) it seems that enzymes with
higher catalytic power are accessible were they selectively
preferred. Thus, if faster and slower enzymes are accessible by
small changes in structure, and kinetic properties reflect
adaptation, the implication is that these properties are
optimized.

Similar evidence of natural selection finely tuning the behavior
of enzymes come from studies of Powers and his coworkers on the
lactate dehydrogenases from the fish <u>Fundulus</u> <u>heteroclitus</u>.
(67). The enzyme is polymorphic, and the two forms have
different temperature optima. Consistent with this variation
being adaptive, the relative abundance of the two forms
correlates closely with the temperature of the water in which the
fish is found. Likewise, the temperature optima of lactate
dehydrogenases from cow and fish correlate with the temperature
in the environment where the enzyme was adapted.(68) The reader
is refered to other examples where similar arguments have been
developed with detailed study.(69-73)

We have recently used site directed mutagenesis to directly
demonstrate a correlation between kinetic variability and
selectability in yeast alcohol dehydrogenase. Mutants of yeast
alcohol dehydrogenase that barely alter the properties of the
protein (either kinetic or physical), when returned to yeast
lacking wild-type alcohol dehydrogenase genes, have been found to
alter the rate of growth of the yeast up to 50%.(65,74)
Although the environment in which the yeast were grown was
artificial, the mutant genes were not chromosomal, and the rate
of growth is highly sensitive to precise conditions,
perturbations in behavior corresponding to energetic differences
of less than 0.5 kcal/mol in a single protein clearly can have

substantial impact on the rate of growth.

Even "semi-random" mutagenesis (defined as mutagenesis not deliberately designed to examine evolutionary issues) is suggestive. Given a set of random mutations with kinetic parameters that have been measured in detail, the distribution of the around a maximum can suggest conclusions about the nature of the optimization process that has produced the enzyme available for study. Consider a kinetic parameter such as k_{cat}. If k_{cat} is maximized, the distribution of k_{cat} values of point mutations is expected to be skewed to lower values (Figure 1). The distribution of k_{cat} values of double mutations is expected to be more skewed in the direction of lower values. Indeed, upon the introduction of still more mutations, the distribution of k_{cat} values is ultimately expected to approach that of a randomly constituted protein, most likely a distribution that can be approximated by a simple exponential curve (Figure 1).

In contrast, consider kinetic parameters such as K_M, where the optimum is presumably not a maximum. Here, the K_M values of random mutants are expected to be more evenly distributed around that of the wild-type protein. Strictly, a non-skewed distribution of mutant behaviors could be observed in behaviors maximized subject to constraint, and behaviors that are not selected. For example, k_{cat}/K_M values larger than the second order diffusion rate constants could confer no additional selective advantage over values equal to this constant; in this case, mutants with k_{cat} values higher than those of wild type should be observed. Similarly, unless we know the distribution of a behavior in a protein of random sequence, we cannot distinguish using this approach a behavior that is optimized for some intermediate value and one that is not selected.

Simple models such as this one are based on the assumption that various behaviors (such as k_{cat} and K_M) are not covariant; that in randomly created structures, there is no correlation between different kinetic parameters. Experiment suggests that this assumption is not a bad approximation. For example, Fersht's laboratory has provided a wealth of detailed kinetic data on mutannt aminoacyl-t-RNA synthetases.(12) These mutants were designed to test hypotheses regarding catalytic activity; they were not intended to test evolutionary questions, so may be regarded as "semi-random" from our point of view. As expected

Figure 1

Variation in Behavior in Randomly Generated
Point Mutations can Suggest Whether the Behavior
is Maximized by Natural Selection

A. Distribution of rates of (i) proteins with random sequences
 (ii) proteins derived by point mutation from a wild type protein where
      rate was maximized by natural selection
 (iii) proteins derived by introducing n mutations into wild type protein

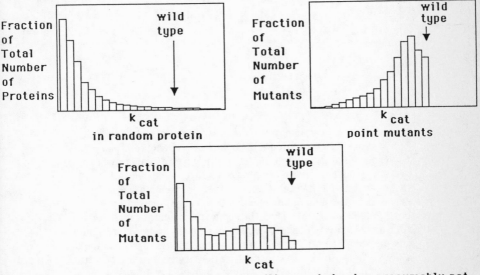

B. Distribution of rates of tyrosine ring flip as a behavior presumably not
   maximized by natural selection

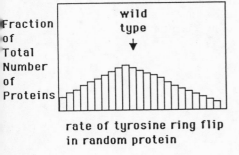

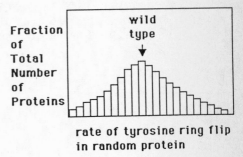

C. Distribution of V/K values of random proteins and
a series of proteins generated by point mutation in a
wild type protein where V/K has been maximized
subject to a constraint, the second order diffusion
rate constant

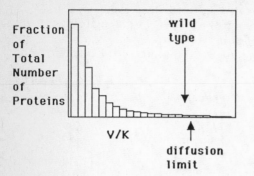

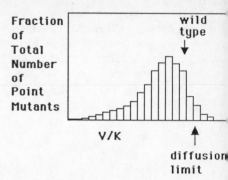

Introduction of random mutations into a protein will create a
set of mutants displaying a distribution in behavior tending
towards the distribution of that behavior in a protein with
random sequence. For example (top pair of diagrams), the
distribution of the rate of catalysis observed in a set of
proteins with random sequence should resemble a simple
decaying exponential. Thus, introduction of point mutations
into a protein with catalytic rate optimized to the right end of
the distribution will create a set of mutants a rate distribution
skewed to the left. In contrast, rate of tyrosine ring flip in
a random polypeptide is distributed around some norm
characteristic for proteins in general. If not selected, this
rate in mutants will be distributed more evenly (2nd pair).
If catalytic rate is maximize with respect to an external
constraint (e.g., V/K and the diffusion limit), the distribution
of this value in mutants shown in pair 3 is expected.

for a maximized trait, the distribution of k_{cat} values of the mutants are skewed towards lower values. Likewise, values of K_M for these mutants are more evenly distributed around the wild type, suggesting that this parameter, if optimized, is not maximized.

A wide range of macromolecular behaviors are widely variable with small perturbations in structure, behave consistent with functional models, and are conserved in widely divergent organisms. Kinetic properties, thermal stability, substrate specificity, dynamic behavior, regulatory phenomena, and genetic structure (including intron structure) all appear to be finely tuned by these criteria. Thus, they should be interpreted functionally, not historically. They reflect natural selection rather than drift or the previous history of the protein.

B. Rather complex behaviors can be created by changing only a few amino acids:

It is important to note that the evolution of new function appears to be rapid under conditions where is can confer selective advantage. In several cases, new catalytic activities are known to have emerged following the introduction of a few point mutations into an existing enzyme as a selective response to an environmental challenge.

Many of these examples are from microorganisms challenged to evolve in new environments.(75-81) When fed butyramide as a primary source of carbon, microorganisms have been found to evolve enzymes capable of hydrolyzing to hydrolyze butyramide as a first step in their degradation.(79) Beta-lactamases act only poorly on cephalosporin. However, when challenged with cephalosporin, enzymes with a single amino acid substitution have evolved that are better able to hydrolyze cephalosporin, and less able to hydrolyze penicillin itself.(80)

Three cases are worth mentioning in further detail. Of 18 mutants in lactose transport proteins in E. coli able to transport maltose, (82) all contained substitutions at either position 177 or 236. These mutants included proteins where Ala-177 had become Val or Thr, and where Tyr-236 had become Phe, Asn, Ser, or His. Variants at position 177 retain ability to transport galactosides, while variants at 236 are defective in transport of galactosides. These results provide an interesting insight into the first steps in the adaptation of this protein. There appear to be only two "dimensions" in structure space (10) along which a lactose carrier can evolve to transport maltose by a single change. In one dimensions, lactose and maltose transport properties are interdependent; in the other, they are independent.

In a second case,(83,84) the gene for beta-galactosidase was deleted from E. coli, and the organism grown with lactose as a carbon source. A new beta-galactosidase then evolved by introduction of 2 point mutations into another (as yet undefined) protein. The detailed kinetics of the intermediate proteins have been studied. The evolution was clearly adaptive and a linear relationship was found between growth rate (with lactose as a carbon source) and the k_{cat}/K_m for hydrolysis of lactose of the evolving gene.(83,84)

Finally, a convincing case is now available for the evolution in just a few years of a new chorismate mutase from an allosteric site on deoxyarabinoheptulosonate phosphate (DAHP) synthase in a laboratory strain of Bacillus subtilis from which the original chorismate mutase had been deleted.(85) The allosteric site on DAHP synthase, the enzyme catalyzing the first step in the biosynthesis of aromatic amino acids, originally bound prephenate, the product at a branch point in the pathway. In its new form, DAHP synthase is now a bifunctional enzyme catalyzing both the first and last step of the pathway. The new enzyme is 50 fold slower than the native enzyme, but still manages to effect a rate enhancement of 4×10^4,

The emergence of a catalytic site from a binding site in such a short time suggests the relative ease of evolution of a new enzyme, provided that a protein performing other functions already exists that contains part of the information needed for assembling the new enzyme. This suggests that biochemical function that might have evolved in this fashion is again better viewed as a product of natural selection, not of the early history of organisms. What other proteins might already contain sufficient information to permit ready evolution of a new catalytic function are difficult to state with certainty. However, species that bind the same cofactor, or the product or substrate of the reaction in some capacity (including proteins catalyzing the formation of the substrate or subsequent reaction of the product) are all candidates.

However, the process of deletion followed by re-emergence of a replacement catalytic function is a general mechanism that can create the appearance of drift in the behavior in proteins. For example, considerable effort has recently been devoted to determining the stereospecificity of the reaction catalyzed by chorismate mutase.(86) If all chorismate mutases display the same stereospecificity, this suggests one of two hypotheses: (a) stereospecificity is not directly selectable, all chorismate mutases are homologous, and stereospecificity is slow to drift because it is structurally coupled to selectable behaviors, or (b) stereospecificity is directly selectable, and the common stereospecificity of all chorismate mutases reflects a combination of convergent evolution and functional constraint on drift. If facile processes lead to the replacement of old chorismate mutases by evolved proteins, this suggests that the fraction of convergent chorismate mutases is higher than in the absence of these processes. This in turn favors explanation (b) over explanation (a).

3. Complicated function appears to have emerged in the modern world by similar processes:

Comparison of homologous proteins indicates remarkable adaptability. a cytochrome P-450 reductase from pig seems homologous, at least in part, with glutathione reductase, ferredoxin reductase, and flavodoxin.(87) Lipoamide dehydrogenase, plasmid-encoded mercuric reductase, and human erythrocyte glutathione reductase are homologous.(88) Carbamoyl-phosphate synthetase may have arisen by the fusion of two genes, one for a glutaminase and one for a synthetase. (89) Enzymes catalyzing consecutive steps in metabolic pathways, including the biosynthesis of methionine (90) appear to have arisen from common ancestors. Ornston and his colleagues have recently suggested that enzymes in the beta-ketoadipate pathway in Pseudomonas putida may be homologous.(91)

Provided information from the fossil record, the time required for these evolutionary events can be estimated. 100 million years appear to be required for two homologous genes to drift to 50% sequence identity provided there are no functional constraints on drift. 500 to 1000 million years is required for sequence identity to be lost entirely. The first number is on the order of time between major geological epochs. The latter number is on

the same order as the time that multicellular life has been
present.

Adaptive Behaviors: Dehydrogenase Stereospecificity

This rather abstract discussion can be illustrated by examining a
single trait that is neither obviously functional nor obviously
non-functional. Reduced nicotinamide adenine dinucleotide
cofactors (NADH and NADPH) bear two hydrogens (Figure 2) at the
4-position, H_r and H_s. The hydrogens are different, and
individual dehydrogenases transfer only one of them. This is a
subtle trait that displays diversity; half of the dehydrogenases
examined transfer H_r, half transfer H_s.(92) (Table 2)

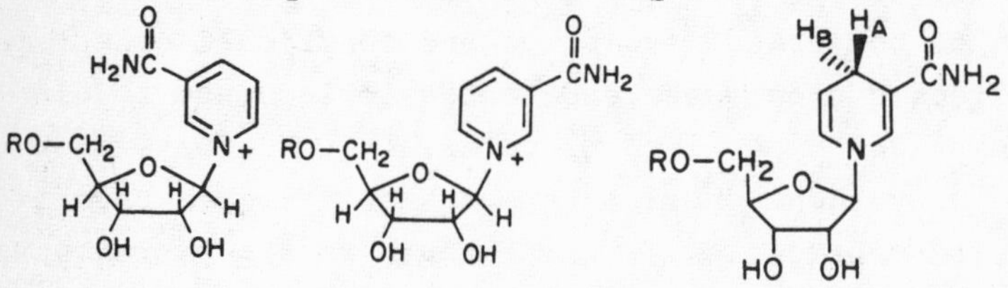

Figure 2: The syn and anti conformers of NADH

Two types of models might explain the stereospecificity of
dehydrogenases dependent on nicotinamide cofactors.(4)
Functional models assume that particular stereospecificities are
identified by natural selection as the ones best suited to help
the host organism survive and reproduce. Historical models deny
a selectable function for stereospecificity. Rather, they view
stereospecificity as a "neutral" trait that is either "drifting"
randomly as the structure of the protein diverges, or is
sufficiently tightly coupled to other functional behaviors in the
protein that it is highly conserved, and thus reflects a randomly
fixed historical accident early in the evolution of a protein.

However, the pattern of stereospecificity in dehydrogenases is
not random. For example, all malate dehydrogenases, including
enzymes from animals, plants, eubacteria, and archaebacteria
transfer H_r.(92) This suggests a priori that this particular
stereospecificity is optimal for this particular reaction. A
functional model based on this suggestion was recently
constructed based on three hypotheses:(93) (a) which hydrogen is
transfered is controlled by the conformation of bound NADH, and
is determined in part by stereoelectronic considerations; (b)
different conformers of NADH have different redox potentials; and

(c) the relative free energies of enzyme-bound species is adjusted in an optimal enzyme to form a "descending staircase" in the direction of the physiological flux.(65)

These arguments (65,93) are published and will not be examined here in detail, except to note that they make a simple prediction: enzymes evolved to reduce unstable carbonyls should have evolved to transfer H_r, while enzymes evolved to reduce stable carbonyls should have evolved to transfer H_s. This prediction is in general confirmed by experiment (Table 2) for dehydrogenases interconverting alcohols and ketones.

Table 2

Stereospecificity of Dehydrogenases Organized by
Redox Potential of Presumed Natural Substrate

log K_{eq}	Enzyme	Stereospecificity
-17.5	Glyoxylate reductase	R
-13.5	Tartronate semialdehyde reductase	R
-12.8	Glycerate dehydrogenase	R
-12.1	Malate dehydrogenase	R
-12.1	Malic enzyme	R
-11.6	L-Lactate dehydrogenase	R
-11.6	D-Lactate dehydrogenase	R
-11.2	Ethanol dehydrogenase	R
-11.1	Glycerol-3-phosphate dehydrogenase	S
-10.9	Homoserine dehydrogenase	S
-10.9	Carnitine dehydrogenase	S
-10.5	3-hydroxyacyl-CoA dehydrogenase	S
- 8.9	3-hydroxybutyrate dehydrogenase	S
- 7.7	Estradiol 17-dehydrogenase	S
- 7.6	3-Oxoacyl ACP Reductase	S
- 7.6	3-Hydroxysteroid dehydrogenase	S

130 Dehydrogenases studied to date, ca. 120 fit correlation, perhaps 10 do not. See ref 4 for discussion

K_{eq}= [NADH][carbonyl][H^+]/[NAD^+][alcohol]

However, it is not obvious from inspection of the crystal structures of dehydrogenases that stereospecificity can be reversed by changing only a few amino acid residues. Further, pairs of homologous enzymes from nature where sequence identity is over 50% always have the same stereospecificity. This raises the possibility that stereospecificity in dehydrogenases might be structurally linked with other behaviors, and therefore not drift even if it were non-functional. Based on this possibility, an alternative historical model can be constructed. Such a model attempts to explain the pattern of stereospecificity purely in terms of pedigree.

Model building must be done carefully, with an eye towards logical rigor. For example, it appears reasonable to many biochemists that the pattern of stereospecificity in dehydrogenases can be understood as resulting from a process where the substrate specificity of an ancestral dehydrogenase can rapidly diverge, but the stereospecificity remain more highly conserved, thus creating a class of enzymes acting on a range of substrates having the same stereospecificity. However, such a model cannot explain the extreme conservation of stereospecificity in malate dehydrogenases, as facile divergence in substrate specificity can in principle create stereochemical diversity in a class of proteins just as easily as can divergence in the mode for cofactor binding.

For example, as discussed above, a deleted enzyme can be replaced by the evolution of a second enzyme (often catalyzing a related reaction). This process provides a mechanism for creating stereochemical diversity in a class of dehydrogenaseses. For example, if a malate dehydrogenase (transferring the pro-R hydrogen) is deleted, and the activity is replaced by the evolution of a 3-hydroxybutyrate dehydrogenase (pro-S specific) with conservation of cofactor stereospecificity, a pro-S specific malate dehydrogenase is created. Because this process is so facile, it is expected except under two circumstances: (a) the deletion of malate dehydrogenase is generally lethal or (b) substrate specificity is too highly conserved for such replacement events to be facile. The first assumption is demonstratably false in many organisms, which can tolerate the deletion of malate dehydrogenase or which lack the enzyme in the wild. Therefore, should we wish to accomodate the facts in Table 2, we must assume that the drift of substrate specificity in alcohol dehydrogenases can only be between structures having similar redox potentials. Thus, the historical model must assume that an active site evolved to handle malate can evolve to handle lactate (a carboxyl group replaced by a hydrogen), but not to handle 3-hydroxybutyrate (a carboxyl group replaced by a methyl) (Figure 3).

Thus, the simplest historical model consistent with the facts presented so far must assume that: (a) there existed at least two ancestral dehydrogenases, one transferring (randomly) H_r, the other transferring (randomly) H_s; (b) the first enzyme had a preference for unstable carbonyl substrates, while the second had a preference for stable carbonyl substrates; and (c) both stereospecificity and substrate type were conserved during divergent evolution, at least to a range of substrates with similar redox potentials.

Once model building is complete placing a historical and a functional model in opposition, we must collect data to distinguish between the two. We summarize relevant facts that have been collected over the past few years::

Fact 1: Stereospecificity in alcohol dehydrogenases correlates with the redox potential of the natural substrate.(1)

Fact 2: The dinucleotide binding domains of the ethanol dehydrogenases from yeast and Drosophila melanogaster appear homologous by sequence comparisons.(4,94) The enzymes catalyze a redox reaction on substrates with intermediary redox potential (Table 2), and the enzymes have opposite stereospecificities with respect to cofactor.(4) Similarly, the dinucleotide binding domains of glyceraldehyde-3-phosphate dehydrogenases and lactate dehydrogenases appear homologous (based on comparisons of their crystal structures).(95) These enzymes also have opposite stereospecificities.

Fact 3: Certain dehydrogenases acting on ethanol and sorbitol are clearly homologous,(94) as are certain ribitol dehydrogenases, glucose dehydrogenasess, and ethanol dehydrogenase.(94)

Fact 4: D and L-lactate dehydrogenases are two classes of enzymes, both catalyzing a reaction with a K_{eq} in the "pro-R region" of the correlation in Table 2, acting on substrates with opposite chiralities, yet having identical stereospecificities with respect to cofactor.(92)

Fact 5: Several pairs of enzymes from widely divergent sources catalyzing a redox reaction far from the break in the correlation (Table 2) share a common stereospecificity. These are perhaps best exemplified by the malate dehydrogenases mentioned above, which include enzymes from eubacteria, archaebacteria, and eukaryotes.(97,98) Similar stereochemical similarities are observed in lactate dehydrogenases.

Fact 6: A pair of dihydrofolate reductases are known that, based on crystallographic data, are not homologous (99) yet have the same stereospecificities. We believe that this is the first clear example of convergent evolution of stereospecificity.

Fact 7: Hydroxymethylglutaryl-CoA reductases from rat and yeast have stereospecificities opposite to that from Acholeplasma.[7](74,100,101)

Fact 8: A survey of dehydrogenases from Acholeplasma and Drosophila suggested that, except for the ethanol dehydrogenase and HMG-CoA reductase mentioned above, dehydrogenases from these two organisms had the same stereospecificities as enzymes from other organisms.(74,98)

These facts are interpreted in terms of functional models quite readily. For example, the correlation suggests that stereospecificity is not strongly selected in enzymes at the "break" in the correlation. The functional model predicts that if stereochemical diversity is observed in analogous enzymes from different organisms, it will be observed in enzymes acting on substrates (e.g. with the general structure CH_2CHO) where the equilibrium constant of the reaction catalyzed is approximately $10^{-11}M$. Both ethanol dehydrogenases and HMG-CoA reductases catalyze reactions consuming such substrates, and both display stereochemical heterogeneity. Malate dehydrogenases, however, acting on substrates well in the pro-R region, natural selection should favor enzymes with pro-R stereospecificity.

But perhaps all enzymes from Drosophila are stereochemically "unusual," or that enzymes acting on substrates with redox potentials far from the break also display stereochemical diversity. The first possibility is ruled out by Fact 8, the second by Fact 2.

Table 3
Recent Results Consistent with Functional Models

Computational verification of stereoselectivity based on
    conformation in reduced nicotinamide cofactors (163)
Crystal structures of dihydronicotinamides (74)
    Correlation between stereoselectivity and redox potential in
nicotinamide model systems (164)
Experimental and theoretical confirmation of arguments regarding
    adjustment of internal equilibrium constants (64)
Convergent stereospecificity of non-homologous dihydrofolate
    reductases (99)
Stereospecificity of $NAD^+/NADP^+$ transhydrogenases (165)

Historical models consistent with these facts must become quite complicated, as they must explain why, in some classes of dehydrogenases, evolutionary processes have not created stereochemical diversity in the time separating archaebacteria, eubacteria, and eukaryotes (even though diversity is known in virtually every other biochemical behavior in this range of organisms), while in organisms that are much more closely related (Acholeplasma, Drosophila, mammals, and yeast) evolutionary processes have produced stereochemical diversity in several classes of enzymes. Further, we must explain why divergence of substrate specificity appears facile (Fact 3), yet has not managed to create stereochemical diversity in malate dehydrogenases, lactate dehydrogenases, glucose-6-phosphate dehydrogenases, and isocitrate dehydrogenases (etc.) by the deletion-replacement process outlined above. Finally, we must explain Fact 1, that stereospecificity within one subclass of dehydrogenases, those interconverting alcohols and ketones, correlates with the redox potential of the natural substrate.

This can be done by a model that assumes multiple ancestral enzymes for HMG-CoA reductases and D- and L-lactate dehydrogenases, single ancestors for malate dehydrogenases and other classes of dehydrogenases where stereochemical uniformity is observed, rapid divergence of stereospecificity, either direct or by replacement events, in ethanol dehydrogenases (but not in other classes of enzymes), and within alcohol dehydrogenases, rapid divergence of substrate specificity as long as the divergence does not convert an enzyme catalyzing a reaction with an equilibrium constant less than 10^{-11} M into one with a K_{eq} greater than 10^{-11} M. Some of these assumptions are more reasonable than others. For example, it is difficult to understand on structural grounds why enzymes that oxidize lactate and malate might be homologous, but enzymes that oxidize malate and 3-hydroxybutyrate cannot be homologous. (Figure 3) Further, in light of the apparent homology between glucose dehydrogenase and ethanol dehydrogenase (Fact 3), essentially any such conservation principles appear to be decidedly ad hoc.

The complexity of this historical model deprives it of both aesthetic and predictive value. However, even unpredictive and unaesthetic models might be correct. For a critical test, we turned to site-directed mutagenesis experiments, and focused on

Figure 3

To Explain the Data in Table 2, the Historical Model Must Assume that the Substrate Specificity of Malate Dehydrogenase <u>Could</u> Drift to Replace a COOH Group by a Hydrogen, but <u>Could Not</u> Drift to Replace a COOH group by a CH_3 Group

malate
all enzymes transfer
H_r

$$HOOC \diagdown CH_2 \diagup C(=O) \diagdown COOH$$

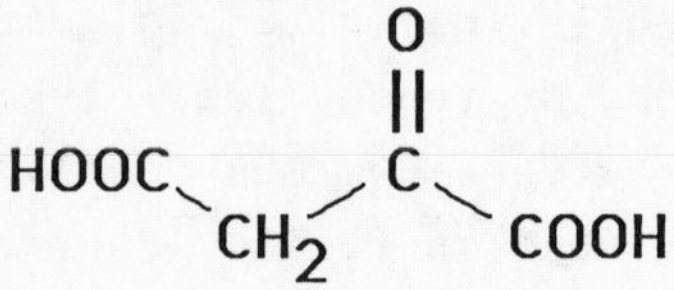

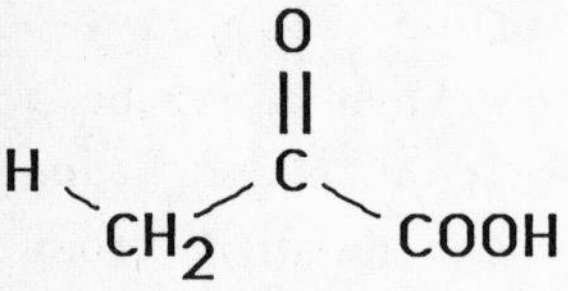

lactate
all enzymes transfer
H_r

$$H \diagdown CH_2 \diagup C(=O) \diagdown COOH$$

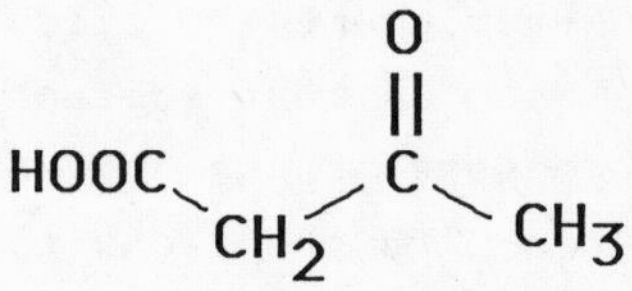

3-hydroxybutyrate
all enzymes transfer
H_s

$$HOOC \diagdown CH_2 \diagup C(=O) \diagdown CH_3$$

the hypothesis central to any historical model that argues that stereospecificity, although serving no selectable function in itself, cannot be reversed without destroying tertiary structure and, with it, catalytic activity. As stereospecificity is determined by the conformation around the glycosidic bond of the cofactor bound in the active site (Figure 4), this means that in alcohol dehydrogenase from yeast or horse (a pair of closely homologous enzymes both transferring H_r), the conformation cannot be reversed by point mutation without destroying catalytic activity.

In alcohol dehydrogenase from horse liver, the syn conformation of NADH (Figure 4) is sterically obstructed by a valine at position 203. Leucine is found in the corresponding position in alcohol dehydrogenase from yeast; here the steric obstruction to the syn conformer is expected to be even larger. However, altering this residue to alanine should remove most of the steric obstruction. While an mutant would be expected to continue to bind NADH predominantly in the anti conformation (the attractive forces binding the carboxamide group remain), the enzyme should become somewhat "stereorandom." If the catalytic activity is diminished in the mutant more than stereospecificity, this result would be consistent with the notion that stereoselectivity is conserved because it is intrically connected with catalytic activity. The result would not rule out the possibility of drift, but it would strengthen the historical view. In contrast, if stereospecificity is decreased in the mutant without a correspondingly large loss in catalytic activity, this would show that stereospecificity was not tightly coupled to tertiary structure and catalytic activity, and suggest that, if not functional, stereospeicifity could drift. Given this result, the identical stereospecificity of malate dehydrogenases must reflect direct function.

The Leu182Ala mutant was made by Arthur Glasfeld using recombinant DNA methods.(74) The mutant was only slightly changed in catalytic activity; k_{cat} was down only by a factor of 3. This itself is a noteworthy result, suggesting a tolerance for structural modification in the active site greater than (we) expected. However, the stereospecificity of the reaction was ten-thousand fold smaller in the mutant than that of the wild type.

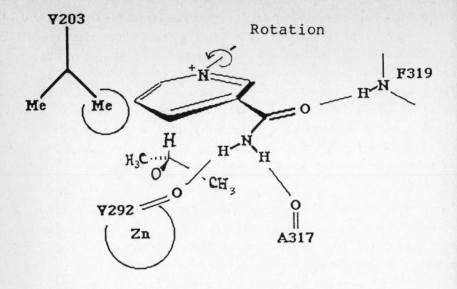

Figure 4: This representation of the active site of horse liver
alcohol dehydrogenase, with the cofactor in an "anti" conformer
on "top" of the substrate, shows that rotation around the
glycosidic bond (marked) to the "syn" conformer would present the
opposite face of the cofactor to the substrate, and thus reverse
stereospecificity of hydrogen transfer. However, the "syn"
conformer is sterically obstructed by the side chain of valine
203 (corresponding to leucine 182 in the alcohol dehydrogenase
from yeast). Replacing the amino acid at this position by
alanine is expected to permit the cofactor to bind in the "syn"
conformation, although it does not remove the "attractive" forces
(the hydrogen bonds to the main chain atoms of phenylalanine 317
and valine 292) which stabilize the "anti" conformer.

The most direct conclusion that can be drawn from these results
is that the apparent diminished stereospecificity of Leu182Ala
arises because replacing Leu by Ala at position 182 removes the
steric obstruction that prevents (in the wild type) the cofactor
from binding in the conformation that presents the pro-S hydrogen
of NADH to the substrate. This removes the obstacle to pro-S
transfer; the hydrogen bonding to the carboxamide group favoring
the "anti" conformer was not deliberately altered, and presumably
is responsible for the stereospecificity in the mutant that
remains.

A variety of control experiments were run to be certain that the
appearance of stereorandomness was not an artifact. For
example, the "stereo-wrong" hydride transfer was inhibited by
pyrazole parallel to inhibition of the activity of the enzyme.
This, together with the fact that the "stereo-wrong" transfer was
not observed in a control containing wild type ADH expressed and
purified in a parallel experiment, argues strongly that the
result was not due to a contaminant of a trace "NAD reductase"
activity with pro-S stereospecificity. Further, the loss in the
stereo-wrong transfer occurred in parallel with inactivation of
the enzyme upon standing.

These results are surprising in several ways. First, judging from
other mutagenesis experiments, the Leu182Ala mutant might have
been expected to be catalytically inactive. For example, a more
remote change in the active site of aspartate aminotransferase
causes a 100,000 fold decline in catalytic activity.(102) If
the loss of catalytic activity had been equal to or greater than
the loss of stereospecificity of the enzyme, this result would be
consistent with the notion that stereospecificity is intrically
connected with catalytic activity. The result would not rule out
the possibility of drift, but it would strengthen the historical
view. Thus, the fact that the catalytic power of mutant
Leu182Ala is lower than WT only by a factor of 3.6 is itself
noteworthy, as it suggests a tolerance for structural
modification in the active site greater than (we) expected.

More remarkable is that the stereospecificity of the mutant
appears to be ten thousand fold lower than wild type, even though
"attractive" forces favoring the "anti" conformation in the
active site were not deliberately altered. Although mutations to
remove these "attractive" interactions are required to completely

reverse stereospecificity, this result by itself rules out a strict correlation between stereospecificity and catalytic activity in YADH.

A statement that a large loss of catalytic activity need not accompany a large loss in stereospecificity contradicts the nearly universal belief at the core of most interpretations of enzymatic stereospecificity in the past decade. Indeed, in dehydrogenases, it has been argued on several occasions that it is impossible for stereospecificity with respect to cofactor to be reversed during the course of evolution.(103,104) This result suggests that dehydrogenases can evolve readily from pro-R specificity to pro-S specificity, if stereospecificity is not a directly selectable trait. This suggestion contradicts the notion that stereospecificity is not directly functional, but is nevertheless highly conserved because it is tightly structurally coupled to tertiary structure and catalytic activity. As this notion is critical for historical models attempting to explain the uniform stereospecificities seen in most classes of dehydrogenases acting on a single substrate (vide supra), these suggestion now appear to make historical models unacceptable as explanations for the stereospecificities of dehydrogenases in all but the most limited cases.

Rather, stereospecificity in dehydrogenases displays all three characteristics expected for a selected macromolecular behavior. It is identical in analogous enzymes from widely divergent organisms (including in at least one pair of enzymes that are almost certainly not homologous), it correlates with chemical details of the catalyzed reaction, and appears capable of drifting if not directly functional.

If stereospecificity in dehydrogenases reflects adaptation rather than history, the implication is that other behaviors of similar structural complexity with similar patterns of conservation are also adaptive. However, it is clear that some stereochemical details of enzymatic reactions are not similarly conserved. For example, we have recently shown that enzymes catalyzing the decarboxylation of beta-keto acids can be found that produce retention, inversion, or racemization.(105) While stereospecificity in decarboxylases certainly cannot drift as rapidly as unconstrained kinetic behaviors, it certainly displays far less uniformity than malate dehydrogenases. Here again,

stereospecificity in malate dehydrogenases appears to be
adaptive, while stereospecificity in beta-keto acid
decarboxylases appears not to be. Based on arguments similar to
these, many enzymatic behaviors can be tentatively assign as
"selectable" or "non-selectable" (Table 4).

Table 4

Candidates for Selectable Macromolecular Traits

Stereoselection between diastereomeric transition states
    NADH-dependent redox reactions
    Phosphoryl transfer reactions
Internal Equilibrium Constants
Kinetic parameters $\pm$ 5%
Stabilty/instability
Substrate specificity against compounds present physiologically

Candidates for Neutral Macromolecular Traits

Stereoselection between enantiomeric transition states
    Decarboxylation of beta-keto acids
    Pyridoxal-dependent decarboxylation of amino acids
Non-equilibrium dynamic motion of proteins
Substrate specificity against compounds not present physiologically

Drifting Structures: Ribonuclease

Site directed mutagenesis is useful in examining behaviors that
are potentially drifting. For example, consider the
ribonucleases (RNases) from bovine, swamp buffalo, and river
buffalo. The three RNases contain 124 amino acids in sequences
that are identical except in six positions (Table 5). The non-
identity could be viewed as an example of neutral drift in
structure. Alternatively, it can be viewed as adaptive, with
RNases with slightly different structures best suited for the
slightly different environments of the swamp, the river, and the
domestic pasture. Each position has its merits. The RNases,
like the lysozymes discussed above, are digestive enzymes that
play a special role in ruminants. They emerged quite recently
together with ruminant digestion, and rapidly evolved during a
period when ruminants themselves diversified into nearly 200
species.

Table 5
Constructing the Sequence of the Ancestor of
Swamp Buffalo, River Buffalo, and Cow Using
The Rule of Maximum Parsimony

Position	Swamp Buffalo	River Buffalo	Cow	Eland	Nilgai	Ancestor
9	Gln	Gln	Glu	Glu	Glu	Glu
19	Ser	Ser	Ala	Ser	Ser	Ser
34	Ser	Asn	Asn	Asp	Ser	Asn
35	Met	Met	Lys	Met	Met	Met
37	Ser	Ser	Lys	Lys	Gln	Lys
65	Lys	Glu	Lys	Lys	Lys	Lys

Corresponding Tree

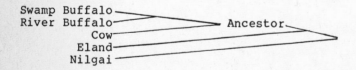

Swamp Buffalo
River Buffalo — Ancestor
Cow
Eland
Nilgai

If we could obtain and examine the intermediates in the divergent
evolution of RNase, we might be able to obtain some clues
regarding the behavioral variation that is associated with this
structural evolution. These intermediates are, of course, the
RNases that were present in the ancestors of modern ruminants.
Regretably, these ancestors are now extinct. However, by

applying a "rule maximum parsimony," we can deduce the sequences of these ancient proteins using information constained in the sequences of their descendents. The rule works as follows.(18) The sequence of the RNase that was the ancestor of the RNases in swamp buffalo, river buffalo, and cow is presumed to be identical to that of the descendents in those parts where the sequences of the descendents are identical. Where the amino acid substitution is different in the descendents, the amino acid is assigned in the ancestral RNase so that smallest number of mutations is needed to create the modern structural diversity.   In making this assignments, sequences of RNases from other sources (here, eland and nilgai) are considered. Using this method, Beintema and his coworkers have assigned a rather complete tree to RNases from modern ruminants.(18)

We have now synthesized the ancestor RNase and certain of its descendents using recombinant DNA techniques. Here again, the work begins with a gene for RNase, here made by chemical synthesis, that is cloned and expressed.(106) Variants of the gene are made by mutagenesis, and the mutant proteins are isolated and characterized.

We have only begun to retrace the evolutionary history of RNase, so our conclusions must be limited. However, from simple kinetic characterization, it appears as if the structural variation represented by the tree above reflects drift, not adaptation.(107) We do not know how this picture will change as we penetrate farther into the evolutionary past. However, in view of the wide variation in properties of modern RNases (vide supra), significant variation in the behavior of modern enzymes must emerge at some point.

HISTORY, THE RNA WORLD, AND EARLY LIFE

Our picture of biological chemistry has been dominated so far by adaptive and drifting behaviors. It is now necessary to balance the discussion by examining behaviors that are found uniformly in the three biological kingdoms but are nevertheless not adaptive. For example, the use of L-amino acids and D-sugars in proteins and nucleic acids, the nearly universal genetic code, and the ribonculeoside portions of cofactors (such as found in ATP, NAD^+, FAD, S-adenosylmethionine) reflect no obvious selected function, and are better understood in terms of their history.

We list seven peculiarities of modern biochemistry as
illustration:

1. At least three mechanistically different classes of
ribonucleotide reductases exist, one employing B-12 as a cofactor
(well studied in _Lactobacillus leichmannii_), one employing non-
heme iron as a cofactor (found in mammals and _E. coli_), and one
employing manganese (from _Brevibacterium ammoniagenes_).
Archaebacterium may contain a fourth type.(108,109)

2. Several distinct fatty acid synthetases are found in different
organisms. They differ in substrate specificity (some using acyl
carrier protein, some coenzyme A), subunit composition and
arrangement, and stereospecificity.(110)

3. Lysine is biosynthesized by two different pathways, one
proceding via diaminopimelate, the other via ketoadipate.(111)

4. Delta-aminolevulinic acid (the first intermediate in the
biosynthesis of porphyrins) is synthesized by two distinct
processes. In one succinyl-CoA and glycine are condensed by an
enzyme requiring pyridoxal phosphate. In the other, glutamate is
reduced and the product rearranged; a structurally complex RNA
molecule is required as a cofactor.(108,112-116) The first
pathway is generally used for the synthesis of corrins, the
second for the synthesis of chlorophyll. However, the second
pathway appears to be used for the synthesis of corrins in some
archaebacteria.

5. Enzymes decarboxylating amino acids are found in two
mechanistically distinct classes. One uses pyridoxal phosphate as
cofactor, the other a N-terminal pyruvoyl residue arising from
post-translationally modified serine. The enzymes display
different stereospecificities. (108,117,118)

6. Two distinct pathways exist for the biosynthesis of
nicotinamide, one involving oxygen, the other not.(119)

7. Membrane constituents are widely different in different
organisms. Animals incorporate fatty acids and steroids.
Eubacteria commonly incorporate hopanoids, including a hopanoid
covalently bound to an RNA moiety. Archaebacterial membranes are
commonly made from terpenoids .(120)

Although functional models have been proposed for various of these
peculiarities,(121) none have been particularly satisfactory.

However, in such complex metabolic properties, history is expected to play a significant role, and our goal is to construct a historical model to explain these and other facts in a general way. Such a historical model must be ad hoc. However, we will be successful if the model devised is internally consistent, chemically plausible, and intrically interconnected with a wide range of modern biochemical phenomena.

The Theoretical Constraint: The RNA World

We begin by assuming that life on earth passed through three episodes (Figure 5).(122-131) In the first (the "RNA world" (123)), catalytic RNA molecules ("ribozymes") were the only genetically encoded components of biological catalysts. The first ribozyme was (presumably) an RNA molecule that catalyzed the RNA-directed polymerization of RNA.

The beginning of the second episode was marked by a "breathrough" to translation, the first synthesis of a protein by the translation of a genetically encoded RNA message in a "breakthrough organism." (122) The breakthrough organism is also the first organsm to contain a ribosome, and only RNA molecules were genetically encoded components of this ribosome. Following the breakthrough, both proteins and RNA molecules served catalytic roles (as they do in the modern world).

The third episode is characterized by the divergent evolution of a "progenote," the most recent common ancestor of the three modern kingdoms of life, archaebacteria, eubacteria, and eukaryotes. Because certain components of modern ribosomes are homologous in all forms of life, the progenote must have lived after (or contemporaneously) the breakthrough. Descendents of the progenote divergently evolved to give the three modern kingdoms of life, archaebacteria, eubacteria, and eukaryotes. (124)

The third episode is distinct because its development can be reconstructed (at least in principle) by direct examination of modern biochemistry (vide infra). However, during both the second and third episodes, proteins evolved to replace at least one (RNA polymerase) and perhaps many ribozymes as catalysts. The goal of model building that begins with these assumptions is to define in as much detail as possible the biochemical development in each episode.

Elements of this model have been discussed for some time. The notion that the earliest form of life contained catalytic RNA was

FIRST ORGANISM

Contained an RNA–directed RNA polymerase that
was an RNA molecule, and no other genetically
encoded catalytic molecules

RNA WORLD

Simple extrapolation from modern
biochemistry is impossible;
Deductions based on organic
chemistry and the deduced metabolis
of the breakthrough organism

BREAKTHROUGH ORGANISM

First organism to synthesize proteins by translation
First organism with genetically encoded message
Complex metabolism, including reactions dependent
on NADH, FAD, coenzyme A, S-adenosylmethionine, ATP
All of the genetically encoded portions of the
catalysts are RNA molecules

Extrapolation from biochemistry
of progenote, together with
assumptions inherent in the
RNA-world model

PROGENOTE

Most recent common ancestor of modern life forms
Existence after the breakthrough secure, as all
ribosomes from modern organisms are homologous

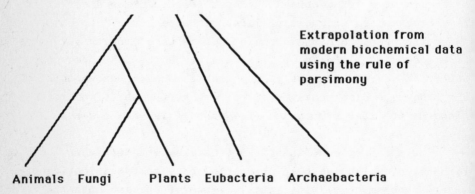

Extrapolation from
modern biochemical data
using the rule of
parsimony

Animals Fungi   Plants Eubacteria Archaebacteria

Figure 5: An outline for the origin of modern life.

proposed by Woese,(125) Crick,(126) and Orgel (127) in the late 1960's. The model was developed in the 1970's, when it was considered respectable enough to be contained in undergraduate textbooks.(128) Nevertheless, RNA was considered to be too poor a catalyst for the RNA world to have lasted for long.

In 1976, Usher and his coworkers discovered the first example of catalytic RNA.(129) Further, significance of the structure of many "ribo-cofactors," first mentioned by Orgel, was appreciated and developed by White.(130) Finally, in what might be regarded as the apogee of theoretical development of this model to date, Visser and Kellogg provided the first clear interpretation of the reactivity and structure of various cofactors in terms of this model.(131)

The model solves four specific problems presented by the biochemistry of modern organisms: (a) The "chicken or egg" problem, originating in the simultaneous need for DNA to make proteins, and proteins to make DNA; (b) The intermediacy of m-RNA between DNA and proteins; (c) The presence of r-RNA as the principal component of ribosomes; (d) The fact that many cofactors contain RNA-like moieties that do not participate in enzymatic reactions.

The model first became popular among molecular biologists after Cech, Altman, and their coworkers demonstrated that catalytically active RNA molecules functioned in the removal of introns from messages and in the processing of transfer RNA molecules. (132,133) A flurry of models describing the origin of life, the RNA world, and the origin of translation based on genetic structure have now emerged.(134-136) As these models largely overlook the chemical and metabolic details of modern organisms (and in most cases do not explain them), we shall not discuss them here except to note their general interest.

For this historical model to explain the four points mentioned above, it must include other assumptions. For example, if the model is to explain the structure of NAD^+ as a vestige of an RNA world, it must assume that the "breakthrough ribo-organism," the first organism to synthesize proteins by translation, had NAD^+ and ribozymes that used NAD^+ as a substrate.

This conclusion follows in part from the chemistry of NAD^+. The RNA part of the cofactor is superfluous with respect to the reactivity of the cofactor. Indeed, for many ribo-cofactors

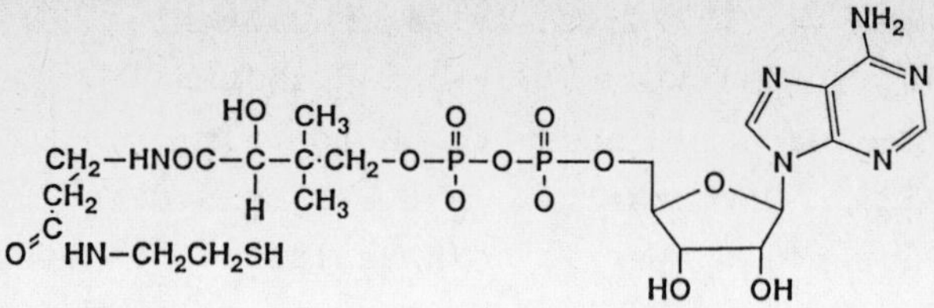

Coenzyme A

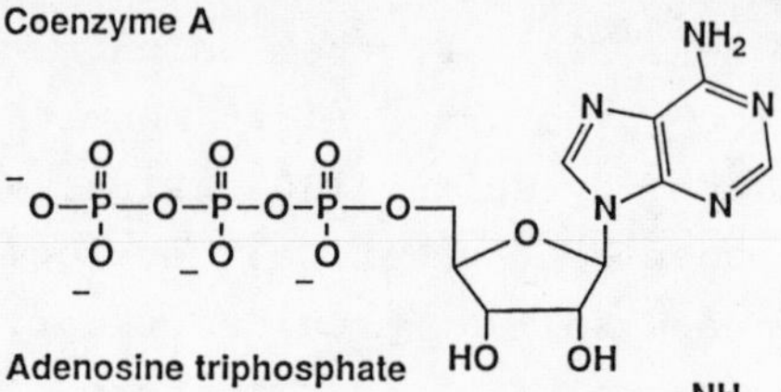

Adenosine triphosphate

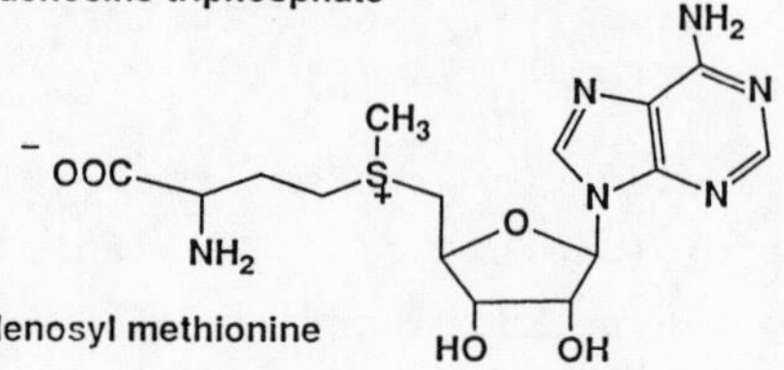

S-adenosyl methionine

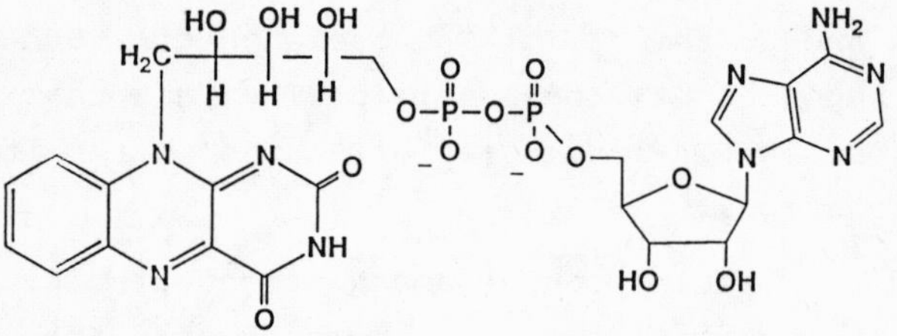

Flavin adenine dinucleotide

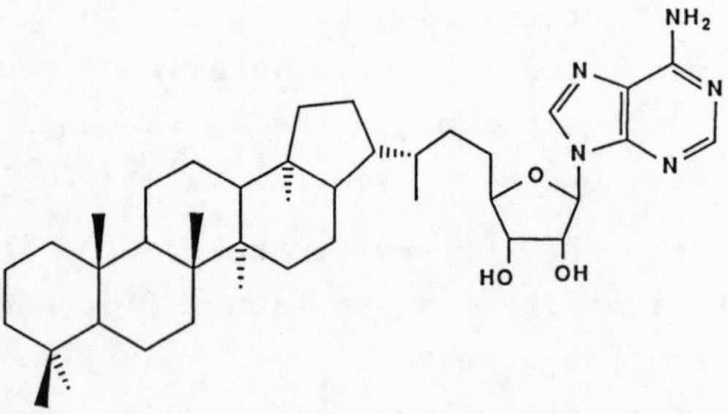

30-(5'-adenosyl)hopane Rhodopseudomonas acidophila

Figure 6. Many cofactors and other biological molecules contain RNA segments that appear to be incidental to their chemical function.

(Figure 6), analogous molecules with similar reactivities perform
identical biological functions without any RNA component. S,S-
dimethylacetate (in an approapriately evolved enzyme) is as
potent a methyl donor as S-adenosylmethionine. (137)
Pyrophosphate is kinetically as competent a phosphorylating agent
as ATP.(138)

The chemical "superfluousness" of the RNA moieties of
ribonculeotide cofactors argues that these structures must have
arisen following the breakthrough. Cofactors emerging in an RNA
world plausibly would use RNA moieties as "anchors" to ribozymes.
Cofactors emerging subsequent to the breakthrough would not;
indeed, they would be expected to be streamlined to reflect the
catalytic power of proteins. Thus, biotin and S,S-
dimethylthioacetate display chemistry expected for cofactors that
arose in the protein world. This point, so eloquently stated 10
years ago by Visser and Kellogg,(131) remains unappreciated by
several groups building models of the RNA world.(139)

In fact, the historical model must assume more than just a single
ribodehydrogenase in the breakthrough organism; in fact, it must
assume that the breakthrough organism had several
ribodehydrogenases interconnected in a rather complex ribo-
metabolism. Otherwise, the model cannot easily explain why the
ribo-cofactor structure was conserved across the breakthrough.
If only one ribo-dehydrogenase existed, and if the metabolism of
the breakthrough organism was primitive, there would be no reason
to preserve the superfluous ribo-structures in the newly evolving
protein world. The new world would have created a new cofactor.
The conservation of the superfluous structural details is best
explained as a result of metabolic coupling between ribozymes and protein
enzymes in the period where both catalyzed redox reactions side
by side.(140), The existence of multiple users therefore
constrained the drift of the RNA structures.(10)

Similar arguments apply for S-adenosylmethionine (implying that
the breakthrough organism had ribotransmethylases), flavin
adenine dinucleotide (implying ribo-oxidases and, perhaps, the
presence of oxygen at the time of the breakthrough), adenosine
triphosphate (implying ribo-phosphate metabolism), ribo-
terpenoids (implying a ribozyme-based isoprenoid chemistry),
coenzyme A (implying ribozyme-catalyzed Claisen condensations),
and other ribo-cofactors. Again, the chemically superfluous RNA

portions of these cofactors are not expected to have arisen in a world already having proteins as catalysts. Again, for the superfluous structures to be conserved across the breakthrough, there must have been several metabolic functions requiring each in the breakthrough organism.(122)

Thus, for the historical model to explain the structures of ribocofactors, protein translation must have been invented in a metabolically complex "breakthrough organism. One cannot dilute this statement without damaging the explanatory coherence of the model. For example, many current models attempt to introduce protein participation in the RNA world as early as possible to assist RNA as a catalyst. This attempt is apparently based on the assumption that RNA molecules "need" proteins, without which they are ineffective catalysts. While this view can be questioned on chemical grounds,(122) explanations based on this modified model are necessarily weakened. To the extent that models assume the involvement of proteins in the world where ribo-cofactors evolved (which must also be the world where metabolism dependent on ribo-cofactors evolved), the model cannot explain why the non-functional fragments of cofactors are RNA and not (for example) proteinaceous.

Certain modern biochemical behaviors can be assigned to the breakthrough organism once we have assumed that an RNA world existed simply by the force of logic, as contrary assignments would render the model senseless. Such traits are those that can be reliably assigned to the progenote by the rule of parsimony (vide infra), intimately involve RNA, and are unlikely on chemical grounds to have emerged in their particular chemical form in a world containing protein catalysts. Ribosomal and transfer RNA, nicotinamide adenine dinucleotide, flavin adenine dinucleotide, and other "ribo-cofactors," and elements of the genetic code are treated for the purpose of this discussion as "canonical" derivatives of the RNA world. While other modern traits obviously might be included in the "canon" given appropriate arguments, we shall focus on these as paradigms that can serve as logical "controls."

Other traits that appear to be nearly universally regarded as remnants of the RNA world (self-splicing, intron structure, 3'-tRNA structures, and non-coding information)(134,135) are not considered here to be canonical. As they are not present in all

kingdoms, the rule of parsimony does cannot reliably place them
in the progenote (let alone the RNA world). In most cases, the
structures clearly have appeared and disappeared in recent
evolution. In several cases, strong arguments can be made that
the traits are functional (vide supra), and could have evolved
convergently. Models of early forms of life that assume that
these traits are primitive lack molecular evolutionary support.
Of course, they might obtain support from arguments of other
types.

Equally important are statements about what we cannot say if we
assume the general outline in Figure 5. The structures of
ancient organisms no more ancient than the most recent common
ancestor can be directly inferred from the biochemical traits of
modern organisms. Thus, simple extrapolation of modern
biochemical behavior can provide insight into the biochemistry of
the progenote (at worst), and the breakthrough organism (at best)
(Figure 5). Thus, metabolic descriptions of these two organisms
are prerequisites for speculations about the biochemistry of
earlier forms of life.

Further, given our limited ability to extrapolate from
modern biochemical structure to the structure of forms of life
earlier than the breakthrough organism, persuasive arguments for
the metabolism and evolution of this and earlier forms of life
must be grounded on facts concerning chemical reactivity, modern
metabolism, and biological plausibility as well as genetic
structure.

Assigning Primitive Traits To Ancient Organisms

To construct models for the metabolism of ancient
organisms, we must distinguish between "primitive" and "derived"
biochemical traits in modern organisms. A primitive trait is one
that was present in the progenote (or earlier). A derived trait
is one that arose more recently.

Working backwards: The rule of parsimony:
A direct and logically consistent approach for assigning
primitive traits involves a comparison of the traits of modern
organisms using the "rule of maximum parsimony." As with
the sequences of ancient RNases (vide supra), the metabolic
pathways of ancient organisms are modeled by a tree that explains
the modern biochemical diversity by postulating the smallest

number of historical events.

The rule can be applied more or less stringently. Traits found
in all three kingdoms (such as NADH) are securely assigned to the
progenote. Traits assigned to the progenote simply because they
are found in two of the derived kingdoms are less secure.
However, if the trait in the modern world serves a selectable
function, and if that function reflects the intrinsic chemical
reactivity of the biological molecules involved, it might have
arisen in the three kingdoms by convergent evolution. Thus, non-
functional traits, and traits not reflecting the intrinsic
chemistry of the moleucles involved, if found in all three
kingdoms, are more reliably assigned to the progenote than
analogous traits that perform selectable functions that reflect
intrinsic chemistry. This latter point can be extremely subtle;
what traits are chemically "arbitrary" (that is, do not represent
chemically efficient solution to a particular biological problem)
depends on pictures of chemical reactivity that are often quite
complex.

Further, parsimony works only if organisms of different species
do not exchange genetic information after they have diverged.
Unfortunately, such "lateral transfer" appears to occur with
unknown frequency even between kingdoms (141-144) The
possiblity of lateral transfer introduces further uncertainty
into assignments of primitive traits based on a comparison of a
small number of organism. Lateral transfer of a trait between
kingdoms would mistakenly suggest that the trait was present in
the progenote. In the context of this discussion, we must simply
recognize that this uncertainty exists and that it may undermine
assignments of traits in specific cases without undermining the
enterprise as a whole.

Because of these caveats, we use here (rather arbitrarily) the
most stringent criterion for assigning traits to the progenote,
that they be present in all three kingdoms, and that they either
be (at least in part) chemically arbitrary (that is, they are not
chemically unique solutions to particular biological problems) or
contain particularly high information content.

Arguments based on structure:
If RNA is structurally present in a metabolic process, this fact
can be used to assign the process to the breakthrough organism.
Here again, the RNA structure cannot be involved in a fashion

that reflects its intrinsic chemical properties or adaptive
function, as such roles could be the result of adaptive evolution
in either the RNA or protein worlds. For example, RNA is
intimately involved in the first step in the biosynthesis of
chlorophyll in a transformation that, from the point of view of
reactivity, could be (and is in other organisms) handled more
easily in other ways.(116,117) The presence of RNA in
ribonuclease P is a similar example where RNA is involved in a
chemical reaction that could (and in other systems is) be done
without RNA.(133)

In contrast, t-RNA molecules are essential components of the
ubiquitin-dependent proteolytic system.(144a) While this might be
interpreted as evidence that the ubiquitin system arose in the
RNA world, it is also clear that the activation of ubiquitin by
uncharged t-RNA molecules makes good metabolic sense: uncharged
t-RNA implies a deficiency of amino acids, for which degradation
of proteins via the ubiquitin-dependent process appears to be a
biologically adapted remedy.

Likewise, self-splicing of introns can be viewed as a sensible
adaptation in the protein world in an environment where introns
occur. It takes advantage of intrinsic reactivity of RNA, the
relative facility of catalyzing transesterifications at
phosphorus, and (at least in principle) simplifies splicing. This
view does not rule out the possibility that self-splicing RNA is
a primitive trait. However, if self-splicing can be easily
viewed as a derived trait, models built based on an assumption
that self-splicing is a "molecular fossil" of the RNA world have
an unreliable foundations.

Evolution Of Catalytic Activity

The origin of a replacement catalytic activity:
Evidently, most ribozymes were replaced by protein enzymes
following the breakthrough. Similar replacement processes are
known in modern evolution, suggesting how they might have
occurred in earlier forms of life (vide supra). Most commonly, a
gene for an enzyme is first deleted. This places selective
pressure on a second biological macromolecule that catalyzes the
deleted process poorly, and the structure of the second
macromolecule then evolved rapidly to create a new enzyme
(possibly not homologous to the first) that fill the role of the

deleted protein. Often, the new protein previously filled a related biochemical role, although the relationship can be chemically quite distant.

Following the breakthrough, the deletion of a ribozyme that catalyzes a metabolic function is a process that would allow natural selection to identify proteins that perform the same step poorly, and permit them to evolve to replace the deleted ribozyme. Given indefinitely long periods of time, ribozymes would be replaced in every case where a protein could catalyze the same reaction with superior selective efficiency.

However, the rate of replacement should be different for different categories of processes and in different organisms. If the deletion is lethal under all environments, or if proteins performing chemically analogous functions are not available, replacement is expected to be slow. In contrast, if the deletion simply implies that a biochemical must be obtained in the diet, or if enzymes catalyzing chemically related processes are already present performing other functions, replacement is expected to be frequent.

For example, the evolved beta-galactosidase (<u>vide</u> <u>supra</u>) arises after only two point mutations of a protein performing another task. The deletion is detrimental to survival only when lactose is the sole carbon source. In contrast, the deletion of a ribosomal RNA molecule is expected to be lethal in the protein world in all environments. Further, the deletion cannot be compensated by a special diet, nor are protein molecules performing chemically analogous reactions likely to be available to step in to rapidly replace ribosomal function. Thus, proteins replacing ribosomal RNA should be extremely difficult to evolve. At best one expects the addition of proteins to an RNA core in the ribosome, which is what appears happened.

RNase P, which catalyzes site-specific hydrolysis important in the maturation of t-RNA, is in between. Deletion is expected to be lethal, and the deleted function cannot be replaced by diet. However, in a well evolved protein world, protenaceous RNases are expected to have evolved to perform other functions. One might expect these to be structurally only a short distance from containing the information necessary to act as site-specific endoribonucleases.

Likewise, different organisms in the modern world appear to
tolerate differently deletions of information. For example,
plants appear to tolerate deletions less well than animals,
judging by their more complete metabolism . Animals are more
self-sufficient than parasites such as trypanosomes, which are in
turn more self-sufficient that the ultimate parasite, viruses.
As a working hypothesis, it is tempting to argue on these grounds
that plants are more likely to contain vestiges of earlier forms
of life than animals, and animals more than viuses.

Working forwards:

Rules of parsimony cannot apply to organisms more recent than the
most recent common ancestor of modern organisms. Thus, our
models of organisms before the breakthrough must be based on
organic chemistry more than biology. Work on chemical reactivity
under "prebiotic" conditions can complement in part limitations
arising from the evolutionary bottleneck represented by the
breakthrough organism. For example, recent work of Eschenmoser
and his coworkers has revolutionized our view of vitamin B-12;
instead of being so "complex" as to be possible only under the
most sophisticated forms of biochemical catalysis, the chemistry
of B-12 appears in many details to be attainable under abiotic
conditions.(145) Likewise, experiments pioneered by Miller,
Ferris, and others have provided much insight into the molecules
that might have been formed abiotically, and therefore available
to early forms of life.(146)

Origin of new catalytic functions:

To model evolution in the RNA world, we must
be concerned about the relative facility of evolving new
catalytic functions. Extrapolating from our limited
understanding about the evolution of new catalytic function in
the modern world, three issues are important: (a) How much
information must be assembled in a biological macromolecule
before it can perform a selectable function? (b) Could this
information come from other biological macromolecules that have
previously evolved to serve another function? and (c) what is the
selective force demanding the creation of the new metabolic
process.

The first issue concerns a subjective evaluation based on
chemistry (how easy is it to catalyze a particular reaction?),
and is intuitive to many organic chemists. For example,

catalysis of the hydrolysis of RNA, the decarboxylation of beta-ketoacids, the transfer of phosphoryl groups, and the hydrolysis of esters is "easy" compared to catalysis of the hydrolysis of amides, the decarboxylation of alpha keto-acids, and the transfer of methyl groups.(10) The distinctions can be quantitated by comparing the rates of enzymes catalyzing various processes, or by model studies.

Processes that are easy to catalyze should arise faster than complex processes. The information content that must be assembled in a ribosome before it can positively contribute to survival due to its ability to synthesize proteins is considerably larger than that for conceivable RNA's that would catalyze most metabolic steps (or entire pathways). The direct implication is that complex metabolism preceded translation, and that the ribosome in the breakthrough organism arose by a combination of metabolic ribozymes that served other selected roles in ribo-metabolism.

Indeed, translation machinery appears to be so complex that one feels compelled to search for functions that various components could have performed by themselves. Non-translational synthesis of oligopeptides almost certainly filled selectable roles before the breakthrough. In this view, the involvement of amino acid-RNA conjugates that resemble charged t-RNA molecules in cell wall biosynthesis and chlorophyll biosynthesis would be vestiges of the RNA world.

However, selection pressures in the RNA world are nearly impossible to model at present, as they depend on chemical factors (the contents of the "soup" following the origin of life) and, more importantly, the metabolism of the riboorganisms themselves. Indeed, before we know what ribozymes could be selected in riboorganisms, one must know the metabolism of the organism in which the selection was taking place. In principle, given an accurate composition of the prebiotic soup, one might attempt to piece together the evolution of metabolism in the RNA world one step at a time, errors in the early part of this process will greatly influence subsequent model building. If the time before the origin of life and the breakthrough was long and the metabolism of the breakthrough organism complex, a detailed effort along these lines is hopeless at present.

However, guesses can be made in some cases. The intrinsic

chemical instability of RNA is likely to make DNA selectively
advantageous under all environments. Likewise, any primitive
organism with limited metabolism is expected to rapidly deplete
the resources it encounters; this is certainly true in the modern
world. Thus, if a long time elapsed between the origin of life
and the breakthrough, some form of "ribo-autotrophy" must have
evolved. Finally, selective pressure is greater to develop
biosynthesis of constituent parts of riboorganisms (e.g., bases,
sugars) than components used only catalytically (e.g.,
ribocofactors). Further details require extensive elaboration,
and are presented elsewhere.(147)

Arguments from Biology

Metabolically complex species living in an RNA world should
behave ecologically as similar species in the modern world.
Thus, the existence of ribo-autotrophs implies the existence of
ribo-heterotrophs. Clear advantages accrue to modern organisms
that specialize to adapt to ecological niches; similar advantages
are expected to apply to riboorganisms. Adaptation involving
speciation and genetic isolation of organisms in the modern world
is also expected in the RNA world. Likewise, competition for
resouces between organisms produces extinction in the modern
world; this is also expected in the RNA world.

Thus, in the time required to assemble the information required
for translation, an ecologically complex RNA world is expected to
have developed. The breakthrough must have conferred selectable
advantage on the organism possessing it (this follows logically
from starting assumption and the fact that modern organisms use
proteins primarily for catalysis). Therefore, following the
breakthrough, there must have been extinction of non-translating
organisms overlapping the environment.

RECONSTRUCTING THE EVOLUTION OF LIFE

To illustrate applications of this method, we shall consider
some special topics in biological chemistry. The models outlined
below do not possess the rigor of canon, as they are based on
assignments of traits as primitive based on assumptions regarding
homology and divergence that can be undermined by lateral
transfer, sequence data, or other facts yet to emerge. Thus, in
the discussion, we parenthetically note caveats to the
interpretation of experimental data. Collectively, however, the

discussion suggests a rich set of experimental possibilities, and virtually every modern biochemical pathway, natural product, or enzyme can be discussed in an analgous way.(147)

Did DNA Emerge Before or After Proteins?

DNA is the primary storage molecule for genetic information in the modern world. However, in the first form of life, RNA was used for information storage. Did DNA emerge as the genetic material before or after the breakthrough? With the tools developed above, we can attempt to construct two competing models, one postulating that DNA emerged after the breakthrough, the other that DNA emerged before the breakthrough, and to decide which model is preferable as a working hypothesis.

Modern organisms contain at least three mechanistically distinct ribonucleotide reductases, enzymes that synthesize deoxyribonucleotides from their corresponding ribonucleotides, (vide supra). This suggests (but does not prove, see Table 1) that modern ribonucleotide reductases, are not homologous. This suggests that proteins catalyzing this reaction apparently arose independently at least three (and possibly four) times following the divergence of the progenote.

The mechanistic diversity of ribonucleotide reductases becomes relevant to the origin of DNA if we assume that an organism stores genetic information in DNA if and only if it has a ribonucleotide reductase. This presumption is supported by the fact that ribonucleotide reductases seems to be quite essential for organisms using DNA, essentially no modern organisms obtain deoxynucleosides exclusively from the diet, even those (including some viruses) that are parasitic in most other metabolic pathways. Conversely, in light of the selective advantage conferred by DNA as an information storage molecule, it is difficult to imagine an organism with a ribonucleotide reductase not using it to synthesize DNA for information storage.

DNA arose after the breakthrough

The notion that DNA arose after the breakthrough is supported by the existence of retroviruses and plant viruses that use RNA as genetic material. One is tempted to argue that RNA viruses are vestiges of viruses infecting the progenote, and perhaps vestiges of the RNA world itself. Further, from some perspectives,(136), the machinery involved in the replication of DNA is highly sophisticated, suggesting that originated in a world of advanced protein catalysts.

If DNA arose after the breakthrough, it could have arisen either before or after the progenote. By maximum parsimony, the fact that DNA is used to store information in all three modern kingdoms would suggest that the progenote used DNA as well. Further supporting this notion is the fact that certain enzymes involved in DNA function (e.g., DNA-dependent RNA polymerase) appear to be homologous in all three kingdoms of life.(148) Thus, the progenote must also have contained a ribonucleotide reductase.

Normally, if the progenote contained a ribonucleotide reductase, one would expect that its descendents would all contain homologous ribonucleotide reductases descendent from the progenotic enzyme. As this does not appear to be the case, an explanation is needed.

The progenotic ribonucleotide reductase could have been either an RNA molecule or a protein molecule. In the first case, the mechanistic diversity observed in modern ribonucleotide reductases would reflect replacement of the ribozymal ribonucleotide reductases by protenaceous ribonucleotide reductases following the divergence of the progenote. Alternatively, if the progenote contained a protenaceous ribonucleotide reductase, the metabolic diversity observed in modern ribonucleotide reductases would reflect protein-protein replacement events following the divergence of the progenote.

Neither picture appears to be completely reasonable. If a ribozymal ribonucleotide reductase arose directly following the breakthrough in a world of protein catalysts, it seems difficult to argue that the same ribozyme did _not_ arise in the RNA world, especially given the general complexity of the RNA world (_vide_ _supra_) and the selective advantage of storing information as DNA. (_Pace_, it is conceivable that the progenotic ribonucleotide reductase required a polypeptide cofactor that could only be synthesized by translation.)

Further, it seems unlikely that the progenotic ribonucleotide reductase was a protein, simply because it seems unlikely that a protenaceous ribonculeotide reductases would have been replaced by other proteins so many times. In proteins central to

metabolism that have been studied, homology appears to be the
rule. Occasionally two mechanistic variants are observed; rarely
are more observed. Further, ribonucleotide reductases is
expected to be replaced only infrequently in organisms that use
DNA to store genetic information , as it appears to
be an important protein in organisms that store genetic
information as DNA (vide supra).

A third possibility is that the progenote did not use DNA to
store genetic information and contained no ribonucleotide
reductase, but that both DNA and modern ribonucleotide reductases
originated after the divergence of the progenote. In this
picture, the homologous DNA-dependent RNA polymerases served some
other role in the progenote, and convergently evolved to their
present role in the three kindgoms after the convergent evolution
of DNA. Convergent evolution of this type seems improbable.
However, in view of the selective advantage of DNA, it is
conceivable.

In principle, the distribution of mechanistic types of
ribonucleotide reductases among the three kingdoms might help
distinguish among these possibilities. If a protenaceous
ribonucleotide reductase was present in the progenote, its
descendents could (in principle) be found in isolated branches of
different kingdoms. If no ribonucleotide reductase (or if a
ribozymal ribonucleotide reductase) were present in the
progenote, the distribution of mechanistically different
ribonucleotide reductases should be superimposable upon the
phylogentic tree. Indeed, in the second case, one might find
remnants of a ribozymal ribonucleotide reductase in some
organisms.

In fact, the different mechanistic types do not divide themselves
neatly along phylogenetic lines; some bacteria have an enzyme
that is mechanistically similar to the mammalian enzyme.
However, insufficient numbers of enzymes have been examined,
especially from archaebacteria, and sequence data are available
for only one mechanistic class. Therefore, given the possibility
of lateral transfer (vide supra), information regarding the
distribution of ribonucleotide reductases is insufficient
to argue strongly that the progenote has such an enzyme and, if
so, what its mechanism was.

DNA arose before the breakthrough

From the dialectical perspective, arguments in support of a late emergence of DNA are at best equivocal. RNA viruses, like all vires, are more likely to be recent innovations than ancient ones. Indeed, (vide supra) the adaptability of viruses and their close coupling to the biochemistry of their hosts makes them unreliable as indicators of anything other than their most recent adaptive needs. Further, it is difficult to evaluate objectively the sophistication of DNA replication machinery. However, short oligoribonucleotides are used as primers in the replication of DNA; this might be interpreted as evidence that DNA replication arose in an RNA world.

However, the relative chemical complexity and information content of translation machinery argues that ribonucleotide reductases arose before translation. The information content of a primitive ribosome (ca. 4500 bases of RNA in the ribosomes, plus ca. 700 more for a simplified set of adaptor molecules, plus information in charging enzymes, plus the information in the message itself) is considerably greater than the information content of a primitive ribonucleotide reductase. Thus, on chemical grounds, ribozymes synthesizing deoxyribonucleotides should have emerged before ribosomes.

Indeed, it is difficult to argue that ribonucleotide reductases did not arise in the RNA world, especially if we must argue that they arose three times independently in the modern world. At the very least, proponents of the "DNA-before-proteins" model must argue that the reduction of ribonucleotides is a chemical task that ribozymes are uniquely unsuited to perform.

If DNA emerged before the breakthrough, the breakthrough organism contained a ribozymal ribonucleotide reductase. In this view, the diversity of mechanistic types in ribonucleotide reductases is explained by the assumption that the ribozymal ribonucleotide reductase present in the breakthrough organism was not replaced rapidly following the invention of translation. Thus, the progenote and its immediate descendents used DNA to store genetic information, but the replacement of ribozymal ribonucleotide reductases by protenaceous ribonucleotide reductases occurred in separate lineages after the divergence of modern kingdoms from the progenote. This picture is consistent with the notion that ribonucleotide reductases can only slowly be replaced in organisms relying on DNA as genetic information. This model treats RNA primers used in DNA replication as vestiges of the RNA world.

Critique:

If we argue that the breakthrough organism stored genetic information as RNA, we must assume either (a) that a ribozymal ribonucleotide reductase arose in the protein world, (b) that ribonucleotide reductase was replaced by non-homologous proteins several times in the modern world, or (c) that DNA emerged several times independently in the modern world. All three assumptions are problematic. (a) If a ribozymal ribonucleotide reductase arose just after the breakthrough, it is difficult to believe that it could not have arisen before; (b) replacement events in the modern world appear to be rare in enzymes performing central metabolic functions; thus, modern descendents of a protenaceous ribonucleotide reductase are expected to have one homologous set of enzymes (or perhaps two, but certainly not three or four); (c) other enzymes (such as DNA-dependent RNA polymerases) involved in the functioning of DNA appear to be homologous in all three biological kingdoms,(148) suggesting that DNA was present in the progenote. Thus, the "DNA-after-proteins" model is faced with a dilemma; to explain the diversity observed in modern ribonucleotide reductases, it must make at least one implausible assumption.

Further, the chemically more complex translation machinery is expected to have arisen after the chemically less complex ribonucleotide reductase. Rhetorically, one might argue that an RNA world capable of inventing NADH should have been able to respond to the selective pressures favoring DNA by inventing a ribonucleotide reductase. Thus, in the absence of further information, it is more plausible that DNA arose before translation, and the mechanistic diversity observed in modern ribonucleotide reductases reflects a delay in the replacement of the ribozymal ribonucleotide reductase until after the divergence of the progenote.

This argument rests on several testable assumptions. Sequences of enzymes involved in DNA synthesis, metabolism, and function from archaebacteria, eubacteria, and eukaryotes should be relevant to the argument that DNA was present in the progenote. To the extent that these enzymes are homologous across the three biological kingdoms, DNA as an information storage molecule can be placed firmly in the progenote. Further, model chemistry reproducing the ribonucleotide reduction would allow us to understand better the possibility that catalysts of such reactions arose early in evolution.

Finally, there has been a very recent suggestion that B-12 and non-B-12 ribonucleotide reductases show sequence similarities in the region of the redox-active cysteines. However, it remains unclear as to what conclusions should be drawn from this fact, as this sequence homology could quite easily reflect convergent evolution, given the potential functional importance of this short polypeptide chain. More sequence and biochemical data on various ribonucleotide reductases would be critical in establishing the lack of homology between various types of proteins. Further progress in crystal structure analysis of these enzymes could also prove decisive.(149)

Finally, the ultimate conclusion would be influenced by the conclusion of a discussion over whether vitamin B-12 emerged before or after the breakthrough. Recent work of Kraeutler, Thauer, and others has shown that vitamin B-12 displays heterogeneity in structure.(149a) For example, in certain organisms, corrinoids contain 5-hydroxybenzimidazole, monomethylbenzimidazole, p-hydroxytoluene, and adenine instead of dimethyl benzimidazole. Structural diversity in this cofactor again must be explained in terms of multiple origin or replacement, and raises interesting questions with respect to biosynthesis. Thus, the dimethylbenzimidazole fragment is normally biosynthesized from flavin; modified structures come from other sources. This in turn requires a discussion of the biosynthesis of flavin, which as FAD is a "canonical" RNA world cofactor.

Metabolic Diversity

Metabolic uniformity generally can be interpreted simply as the result of divergent evolution of an enzyme, structure, or pathway originally present in the progenote. Metabolic diversity requires special explanation. In this light, we should conclude by considering possible general explanations for metabolic diversity. Where organisms synthesize the same compound by different chemical pathways, or have enzymes employing different chemical mechanisms, four general explanations are possible.

1. Proteins catalyzed the same pathway in the progenote, but individual enzymes were replaced subsequent to divergence. The replacement could either be "neutral" or adaptive. For example, a decarboxylase without pyridoxal might have evolved to replace a pyridoxal-dependent enzyme in an environment lacking pyridoxal; non-zinc enzymes might evolve to replace zinc-dependent enzymes in the absence of zinc. Intermediates in pathways arising by this process should be the same in different organisms, and enzymes catalyzing most of the steps in the pathway should be homologous. However, protein-protein replacement processes are less likely in "essential" pathways, where the product of the pathway cannot be obtained in the diet, and in plants; replacement processes are more likely given enzymes evolved to catalyze chemically related reactions, where the product of the pathway can be found in the diet, and in animals.

The diversity in mechanism in enzymes oxidizing alcohol (including enzymes dependent on zinc, iron, and those requiring no metal) might be explained by such a process. Ethanol oxidation is found throughout biology and, as a plausible end point for fermentative pathways, the reaction could hardly have been avoided in early metabolism. While limited sequence similarities between zinc and non-zinc enzymes suggests distant homology, and perhaps divergence in the protein world, it is also possible that the diversity reflects protein-protein replacement events with the larger category of NAD^+-dependent dehydrogenases.

2. The pathway was present in the progenote in a protein-catalyzed form, with subsequent replacement of the entire pathway via introduction of new pathways adapted to special circumstances, "pathway capture," and other replacement processes. For example, if an organism contains a pathway that produces a product that is structurally similar to an advanced intermediate on another pathway, the second pathway could be replaced by the first.

The distinct diaminopimelic acid and 2-ketoadipate pathways for the biosynthesis of lysine may have arisen in this way. Lysyl t-RNA's, lysyl aminoacyl-tRNA-synthetases, and lysine codons are homologous in all organisms studied, suggesting that lysine was incorporated by translation into proteins in the progenote. Although it is possible that the progenote obtained lysine in its diet, biosynthesis seems more plausible for any constituent.

Diaminopimelate serves metabolic roles other than as an intermediate in the biosynthesis of lysine, suggesting the possibility of a separate pathway to these compounds that "captured" the biosynthetic pathway for lysine. In this picture, aminoadipate pathway is the more ancient.

3. The pathway was present in the progenote, but in an RNA-catalyzed form. Replacement of the ribozymes by protein enzymes after divergence of the three main kingdoms would produce protein enzymes that need not be homologous catalyzing analogous steps on analogous pathways. (Caveat: if substantial time elapsed between the breakthrough and the divergence of the modern kingdoms, only biochemical pathways that are difficult to replace are likely to fall into this category.)

As discussed above, the two mechanistic classes of ribonucleotide reductases, including B-12 enzymes found in lactobacilli, clostridia, and Euglena, the non-B-12 enzymes found in E. coli mammals and other higher organisms, and the manganese dependent enzymes found in certain bacteria, might best be explained by this process. In general, an argument that the progenote contained an RNA-dependent process implies that the process was also present in the breakthrough organism.

4. The pathway was not present in the progenote in any form, but arose independently in various lineages descendent from the progenote. Here, one expects different pathways to the same product catalyzed by non-homologous proteins. However, certain biosynthetic pathways make more chemical sense than others, and therefore might be expected to arise convergently. For example, serine certainly should be biosynthesized from glyceric acid; identical pathways to serine from glyceric acid is therefore not strong evidence that the pathway existed in the progenote. In contrast, the biosynthesis of histidine from purines makes little chemical sense; if the pathway is conserved, it is a strong indicator of homology.

The mechanistic diversity observed in proteases (including enzymes using serine, cysteine, metals, and carboxylate groups) may be best explained by independent evolution following the divergence of the progenote. Naively, proteases have value only after translation was widespread. However, one might argue that intracellular proteases were needed soon after the breakthrough. By some lines of reasoning, this suggests that relatively little time elapsed between the breakthrough and the progenote.

Other Pathways

We cannot discuss every interesting biochemical behaviors that appear to reflect history more than function. Table 6 addresses a question of particular interest: Was the breakthrough organism photosynthetic? Here again, the discussion does not resolve this question, but it does suggest experimental work. The question is especially intriguing for another reason. The participation of RNA in the biosynthesis of chlorophyll in the modern world (116,117) is consistent with the possibility that the breakthrough organism was photosynthetic (Table 6). The origin of photosynthesis can be approximately dated by the appearance of oxidized sediments in datable geological strata.(150) Fossils of organisms are known preceeding this time. Thus, if the metabolism outlined in Table 7 is correct, these fossils are fossils of ribo-organisms.

CONCLUSION

Constructing functional and historical models for biochemical behaviors requires information from every area of chemistry, biochemistry, and biology. Testing these models requires an equally multidisciplinary effort. We have presented here only one example where questions raised by the models have been resolved (at least partly) satisfactorily. Even here, the progress was made only after many years of experimental work.

But implicit in this discussion is a warning. As difficult as progress is, and as convoluted as the arguments sometimes seem, the bioorganic chemist in the 1980's can no longer safely dismiss the themes developed here as "mere biology," and retreat to the descriptive "hard" natural products chemistry of the past decades. Nothing from biology makes sense except in the light of evolution. This is also true when the "things" from biology are mere chemicals. Without understanding the evolution of biological chemicals, we can never understand their chemistry. Yet without understanding the chemistry of biological molecules, we can never understand biology.

Table 6
Was the Breakthrough Organism Photosynthetic?

YES: The first step in the "C_5" pathway for the biosynthesis of chlorophyll involves RNA intimately in the catalytic step (Schoen et. al. 1986)

NO: But the C_5 pathway is not distributed in modern organisms as one would expect for a primitive trait. Plants have it. But both mammals and bacteria on opposite sides of the tree have an alternative pathway.

YES: But this distribution merely suggests that the enzymes catalyzing the first step of the alternative pathway in bacteria and mammals are not homologous. Bio-organic data suggest that this may be true. The enzyme from <u>Rhodopseudomonas spheroides</u> has a lysine in the active site, while the corresponding enzyme from rat may have an active site cysteine (Nandi, 1978).

AMBIGUITIES:
1. Arguments for an active site cysteine in the mammalian enzyme are not strong
2. At least some bacterial enzymes appear homologous by sequence to mammalian enzymes. However, a cysteine is highly conserved, not a lysine.
3. Some organisms have both pathways

The area is rich with suggestions for new experiments.

Table 7
One Model for the Breakthrough Organism

Reactions Part of the Breakthrough Organism's Metabolism
    Oxidation/reduction reactions, aldol condensations, Claisen
       condensations, transmethylations
    Lived in an aerobic environment, were photosynthetic
    Degraded fatty acids, synthesized terpenes
    Used DNA to store information
    Energy metabolism based on ATP
    Modified RNA bases

Reactions Not Part of the Breakthrough Organism's Metabolism
    Biotin-dependent carboxylations
    Fatty acid synthesis
    Pyridoxal chemistry, transaminations

References

1 Knowles JR (1987) Science 236:1252
2 Lewontin RC (1979) Sci Am 239:156
3 Kimura M (1982) Molecular Evolution Protein Polymorphism and the Neutral Theory, Berlin, Springer-Verlag
4 Benner SA, Nambiar KP, Chambers GK (1985) J Am Chem Soc 107:5513
5 Dequard-Chablat M J Biol Chem 261:4117 (1986)
6 Biswas TK Getz GS (1986) J Biol Chem 261:3928
7 Pietruszko R (1982) Meth Enzymol 89:428
8 Stalker DM, Hiatt WR, Comai L (1985) J Biol Chem 260:4724
9 Both GW, Shi CH, Kilbourne ED (1983) Proc Nat Acad Sci 80:6996
10 Benner SA; Ellington AD (in press) CRC Crit Rev Biochem
11 Miller JH (1979) J Mol Biol 131:249
12 Fersht AR, Leatherbarrow RJ, Wells TC (1987) Biochem 26:6030
13 Shortle D, Lin B (1985) Genetics 110:539
14 Serpersu EH, Shortle D, Mildvan AS (1986) Biochem 25:68
15 Cone JL, Cusumano CL, Taniuchi H, Anfinsen C (1971) J Biol Chem 246:3103
16 Knowles JR (1988) Cold Spr Harbor Symp Quant Biol, 51:
17 Strauss D, Raines R, Kawashima E, Knowles JR, Gilbert W (1985) Proc Nat Acad Sci 82:2272
18 Beintema JJ, Fitch WM, Carsana A (1986) Mol Biol Evol 3 262
19 Capasso S, Giordano F, Mattia CA, Mazzarella L, Zagari A (1983) Biopolymers 22:327
20 Piccoli R, D'Alessio G (1984) J Biol Chem 259:693
21 Vescia S, Tramontano D (1981) Mol Cell Biochem 36:125
22 Levy CC, Karpetsky TP (1981) Enzymes as Drugs, Holcenberg JS, Roberts J eds
23 Palmer KA, Scheraga HA, Riordan JF, Vallee BL (1986) Proc Nat Acad Sci 83:1965
24 Shapiro R, Riordan JF, Vallee BL (1986) Biochem 25:3527
25 Wills C; Joernvall H (1979) Eur J Biochem 99 323
26 Branden CL, Jornvall H, Eklund H, Furugren B (1975) The Enzymes, 3rd ed. 11:182
27 Joernvall H, von Bahr-Lindstroem H, Jany K-D, Ulmer W, Froeschle M (1984) FEBS Lett 165:190
28 Craik CS, Largman C, Fletcher T, Roczniak S, Barr PJ, Fletterick R, Rutter WJ (1985) Science 228:291
29 Kreitman M (1983) Nature 304:412
30 Jacq C, Miller JR, Brownlee GG (1977) Cell 12:109
31 Vanin EF (1984) Biochem Biophys Acta 782:231
32 Miyata T, Hayashida H (1981) Proc Nat Acad Sci 78:5739
33 Li W-H, Wu, Chung-I L Chi-Cheng (1984) J Molec Evol 21:58
34 Li W-H, Gojobori T, Nei M (1981) Nature 292:237
35 Gitlin D, Gitlin JD (1975) The Plasma Proteins Putman FW ed, New York, Academic Press
36 Perler F, Efstradiatis A, Lomedico P, Gilbert W, Koler R, Dodgson J (1980) Cell 20:556
37 Crabtree GR, Comeau CM, Fowlkes DM, Fornace Jr AJ, Malley JD, Kant JA (1985) J Mol Biol 185:1
38 Minghetti PP, Law SW, Dugaiczyk (1985) Mol Biol Evol 2:347
39 Lewontin RC (1985) Ann Rev Genet 19:81
40 Weatherall DJ, Clegg JB (1976) Ann Rev Genet 10:157; Allison AC (1955) Cold Spring Harbor Symp Quant Biol 20:239
41 Yang Y-C, Ciarletta AB, Temple PA, Chung MP, Kovacic S, Witek-Giannotti JS, Leary AC, Kriz R, Donahue RE, Wong GG, Clark SC (1986) Cell 47:3
42 Hill RE, Hastie ND (1987) Nature 326:96
43 Stewart C-B, Schilling JW, Wilson AC (1987) Nature 330:401

44 Dayoff MO (1978) Atlas of Protein Sequence
45 Lewin R (1986) Science 232:578
46 Diamond JM (1986) Nature 321:565
47 Dykhuizen DE (1978) Evolution 32:125
48 Koch AL (1983) J Molec Evol 19:455
49 Li W-H (1984) Mol Biol Evol 1:213
50. Benner SA (in press) Molecular Structure and Energetics, J.F. Liebman and A. Greenberg, eds., Deerfield Beach, Florida, VCH Publishers
51 Domingo E, Sabo D, Taniguchi T, Weissmann C (1978) Cell 13:735
52 Sharp PM, Rogers MS, McConnell DJ (1985) J Molec Evol 21:150
53 Grantham R, Gautier C, Gouy M, Jacobzone M, Mercier R (1981) Nucl Acids Res 9:r43
54 Gouy M, Gautier C (1982) Nucl Acids Res 10:7056
55 Konigsberg W, Godson GN (1983) Proc Nat Acad Sci 80:687
56 Bennetzen J, Hall B (1982) J Biol Chem 257:3026
57 Grosjean H, Fiers W (1982) Gene 18:199
58 Blaisdell BE (1983) J Molec Evol 19:226
59 Rubin CM, Houck CM, Deininger PL, Friedmann T, Schmid CW (1980) Nature 284:372
60 Modiano G, Battistuzzi G, Motulsky AG (1981) Proc Nat Acad Sci 78:1110
61 Lipman DJ, Wilbur WJ (1983) J Mol Biol 163:363
62 Lanave C, Preparata G, Saccone C (1985) J Mol Evol 21:346
63 Benner, SA (submitted) Chem Rev
64 Ellington AD, Benner, SA (1987) J Theor Biol 127:491
65 Ellington AD (1988) Thesis, Harvard University
66 Liang N, Pielak GJ, Mauk AG, Smith M, Hoffman BM (1987) Proc Nat Acad Sci 84:1249
67 Place AR, Powers DA (1979) Proc Nat Acad Sci 76:2354
68 Borgman V, Moon TW (1975) Can J Biochem 53:998
69 Hennessey Jr JP, Siebenaller JF (1987) Biochem Biophys Acta 913:285
70 Koehn RK (1969) Science 163:943
71 Merritt RB (1972) Am Nat 106:173
72 Day TH, Hiller PC, Clarke B (1974) Biochem Genet 11:141
73 Day TH, Needham L (1974) Biochem Genet 11:167
74 Glasfeld A (1988) Thesis, Harvard University
75 Zimmermann T, Kulla HG, Leisinger T (1983) Experientia 39:1429
76 Campbell HH, Lengyel JA, Langridge J (1973) Proc Nat Acad Sci 70:1841
77 Boronat A, Aguilar J (1981) Biochim Biophys Acta 672:98
78 Burleigh BD Jr, Rigby PWJ, Hartley BS (1974) Biochem J 143:341
79 Betz JL, Brown PR, Smyth MJ, Clarke PH (1974) Nature 247:261
80 Hall A, Knowles JR (1976) Nature 264:803
81 Wills C (1976) Nature 261:26
82 Brooker RJ, Wilson TH (1985) Proc Nat Acad Sci 82:3959
83 Murray M, Osborne S, Sinnott ML (1983) J Chem Soc Perkin Trans II 1595
84. Hall BG, Yokoyama S, Calhoun DH (1984) Molec Biol Evol 1:109
85 Llewellyn DJ, Daday A, Smith GD (1983) J Biol Chem 255:2077
86 Grimshaw CE, Sogo SG, Copley SD, Knowles JR (1984) J Am Chem Soc 106:2699
87 Haniu M, Izanagi T, Miller P, Lee TD, Shively JE (1986) Biochem 25:7906
88 Stephens PE, Lewis HM, Darlison MG, Guest JR (1983) Eur J Biochem 135:519
89 Nyunoya H, Broglie KE, Lusty CJ (1985) Proc Nat Acad Sci 82:2244
90 Belfaiza J, Parsot Cl, Martel A, Bouthier de la Tour C, Margarita D, Cohen GN, Saint-Girons I (1986) Proc Nat Acad Sci

83:867
91 Yeh W-K, Fletcher P, Ornston LN (1980) J Biol Chem 255:6342
92 You K-S (1985) CRC Crit Rev Biochem 17:313
93 Benner SA (1982) Experientia 38:633
94 Jornvall H, Persson M, Jeffrey J (1981) Proc Nat Acad Sci
78:4226
95 Rossmann MJ, Liljas A, Braenden C-I (1975) The Enzymes 11:61
96 Garavito RM. Rossmann MG, Argos P, Eventoff W (1977) Biochem
16:5065
97 Goerisch H, Hartl T, Grussebueter W, Stezowski J (1982)
Biochem J 226:885
98 Allemann R, Hung R, Benner SA (in press) J Am Chem Soc 110:
99 Matthews DA, Smith SL, Baccanari DP, Burchall JJ, Oatley SM,
Kraut J (1986) Biochem 25:4194
100 Dugan RE, Porter JW (1971) J Biol Chem 246:5361
101 Beedle AS, Munday KA, Wilton DC (1972) Eur J Biochem 28:151
102 Cronin CN, Malcolm BA, Kirsch, JF (1987) J Am Chem Soc
109:2222
103 Oppenheimer NJ (1986) "Proceedings of the Steenbock
Symposium" Madison Wisconsin July (1985) New York Elsevier (1986)
29-43
104 Garavito RM, Rossmann MG, Argos P, Eventoff W (1977) Biochem
16:5065
105 Piccirilli JA, Benner SA (1987) J Am Chem Soc 109:8084
106 Nambiar KP, Stackhouse J, Stauffer DM, Kennedy WP, Eldredge
JK, Benner SA (1984) Science 222:1299
107 Presnell SR (2988) Thesis, Harvard University
108 Walsh C (1979) Enzymatic Reaction Mechanisms, San Francisco
Freeman
109 Hogenkamp HPC, Follmann H, Thauer RK (1987) FEBS Lett 219:197
110 Nes WR, Nes WD (1980) Lipids in Evolution, NY Plenum Press
111 Vogel HJ (1965) Evolving Genes and Proteins, Bryson V and
Vogel HJ eds, NY, Academic Press
112 Kannangara CG, Gough SP, Oliver RP, Rasmussen SK (1984)
Carlsberg Res Comm 49:417
113 Schoen A, Krupp G, Gough S, Berry-Lowe S, Kannangara CG,
Soell D (1986) Nature 322:281
114 Weinstein JD, Beale SI (1985) Arch Biochem Biophys 239:87
115 Huang D-D, Wang W-Y, Gough SP, Kannangara CG (1984) Science
225:1482
116 Nandi DL (1978) Arch Biochem Biophys 188:266
117 Gerdes HJ, Leistner E (1979) Phytochem 18:771
118 Benner SA, Piccirilli JA (submitted) Top Stereochem
119 Scott TA, Bellion E, Mattey, M (1969) Eur J Biochem 10:318
120 Neunlist S, Rohmer M (1985) Biochem J 228:769
121 Jencks WP (1975) Adv Enzymol 43:219
122 Benner SA, Ellington AD (in press) Nature 329:295; Benner SA
et al., Cold Spring Harbor Sym. Quant. Biol. 51
123 Gilbert W (1986) Nature 319:818
124 Woese CR (1987) Microbiol Rev 51:221
125 Woese CR (1967) The Origins of the Genetic Code, Harper & Row
126 Crick F (1968) J Mol Biol 38:367
127 Orgel LE (1968) J Mol Biol 38:381
128 Lehninger A (1972) Biochemistry, New York, Worth Publishers
129 Usher DA, McHale AH (1976) Proc Nat Acad Sci 73:1149
130 White III HB (1976) J Molec Evol 7:101
131 Visser CM, Kellogg RM (1978) J Mol Evol 11:171
132 Bass BL, Cech TR (1984) Nature 308:820
133 Guerrier-Takada C, Altman S (1983) Science 223:285
134 Darnell JE, Doolittle WF (1986) Proc Natl Acad Sci 83:1271
135 Weiner AM, Maizels N (1987) Proc Nat Acad Sci 84:7383

136 Alberts BM (1986) Amer Zool 26:781-796
137 Cantoni GL (1962) Methods Enzymol 5:743
138 Siu PML,HG Wood (1962) J Biol Chem 237:3044
139 Maizels N, Weiner AM (1987) Nature 330:616
140 Hanson KN, Rose IA (1975) Acc Chem Res 14:1
141 Carlson TA, Chelm BK (1986) Nature 322:568
142 Leunissen JAM, de Jong WW (1986) J Mol Evol 23:250
143 Lewin R (1985) Science 227:1020
144 Syvanen M (1985) J Theor Biol 112:333
144a Ciechanover A, Wolin SL, Steitz JA, Lodish HF (1985) Proc
Natl Acad Sci 82:1341
145 Eschenmoser A (in press) Angew Chem
146 Miller SL (1955) J Am Chem Soc 77:2351
147 Benner SA, Ellington AD (in preparation)
148 Gropp F, Reiter WD, Sentenac A, Zillig W, Schnabel R, Thomm
M, Stetter KO (1986) System Appl Microbiol 7:95
149 Lin A-NI, Ashley GW, Stubbe J (1987) 26:6905
149a Krautler B, Moll J, Thauer RK (1987) Eur J Biochem 162:275
150 Strother PK Barghoorn (1980) Origins of Life and Evolution,
Halvorsen HO, van Holde KE eds, New York, Liss
151 Bradford AP, Aitken A, Beg F, Cook KG, Yeaman SJ (1987)
FEBS Letters 222: 211
152 Frantz B, Chakrabarty AM, (1987) Proc Natl Acad Sci 84:4460
153 Grenett HE, Ledley FD, Reed LL, Woo SLC, (1987) Proc Natl
Acad Sci 84:5530
154 Sikorav J-L, Krejci E, Massoulie J (1987) EMBO J 6:1865
155 Schinina ME, Maffey L, Barra D, Bossa F, Puget K, Michelson
AM (1987) FEBS Letters 221:87
156 Shanske S, Sakoda S, Hermodson MA, DiMauro S, Schon EA (1987)
J Biol Chem 262:14612
157 Parson CA (1987) Proc Natl Acad Sci 84:5207
158 Anton IA, Duncan K, Coggins JR (1987) J Mol Biol 197:367
159 Jornvall H, Hoog J-O, Bahr-Lindstrom von H, Vallee BL (1987)
Proc Natl Acad Sci 84:2580
160 Volokita M, Somerville CR (1987) J Biol Chem 262:15825
161 Pilkis SJ, Lively MO, El-Maghrabi MR (1987) J Biol Chem 262:
162. Hammond GL, Smith CL, Goping IS, Underhill DA, Harley MJ,
Reventos J, Musto NA, Gunsalus GL, Bardin, CW (1987) Proc Natl
Acad Sci 84:5153
163 Wu YD, Houk KN (1987) J Am Chem Soc 109:2226
164 Ohno A, Ohara M, Oka S (1986) J Am Chem Soc 108:6438
165 Kaplan NO (1967) Methods Enzymol 10:317